Concorde: The Inside Story

Geoffrey Knight

Concorde:

The Inside Story

Weidenfeld and Nicolson London

First impression May 1976
Second impression October 1976

Published by George Weidenfeld and Nicolson Limited,
11 St John's Hill, London SW11

ISBN 0 297 77114 0

Printed in Great Britain by
Butler & Tanner Limited
Frome and London

Contents

Illustrations

The above photographs are reproduced by courtesy of British Aircraft Corporation Limited

The jacket photograph is reproduced by courtesy of British Airways

Foreword

I have been greatly assisted in putting together this personal record of the Concorde achievement by the reminiscences and evidence of good friends and colleagues. These include Sir Archibald Russell, British designer of Concorde; Sir Stanley Hooker, 'father' of the Olympus engine and Technical Director of Rolls-Royce (1971) Limited; Mick Wilde, Concorde Project Director for the British Aircraft Corporation, and General Henri Ziegler, formerly President/Director General of Sud Aviation and then of Aérospatiale. My thanks also go to the Rt. Hon. Julian Amery, MP, Minister of Aviation in 1964 and the Rt. Hon. Anthony Wedgwood Benn, MP, Minister of Technology 1967–70, and Secretary of State for Industry 1974–5, who were kind enough to refresh my memory about events during their periods of office.

I have been greatly helped in research work by Mr Norman Barfield of BAC.

We all of us, especially myself, owe a debt of thanks to Mr Brian Connell for helping us to record our recollections and, at the same time, to combine them into a narrative.

Introduction

Concorde at long last is in airline service. On 21 January 1976 an Air France Concorde took off from Paris to fly to Rio via Dakar and, at precisely the same moment, a British Airways Concorde took off from London for Bahrain. Twenty years from its inception, the supersonic transport project has become a reality: Concorde has been granted a Certificate of Airworthiness and is now carrying fare-paying passengers on regular services.

Of course, these first two operations mark only the beginning of what still remains to be achieved. The Bahrain route itself will be only the first stage of an operation which will extend from Bahrain to Singapore and then from Singapore to Melbourne. For this to happen, a supersonic corridor has to be negotiated across the southern part of India.

However, we do at last have the reality of Concorde on the air routes of the world and I have no doubt that from now on its operation will be extended in many directions. I hope that during the next twelve months we shall see this airplane operating to places like Tokyo, Johannesburg, Montreal, Caracas and other important destinations in the world, and I am certain that this will mean, for Air France and for British Airways, the establishment of a position in the market which will be the envy of other airlines the world over.

The major routes which, from the traffic point of view, have captured the imagination of people on both sides of the Atlantic are London–New York and Paris–New York, with – to a lesser but still important extent – London–Dulles

(Washington) and Paris–Dulles. In 1975 British Airways put in its application to operate two Concorde services a day to John F. Kennedy Airport, New York, from London Heathrow, and Air France put in its application to operate a similar number of services from Paris to New York. At the same time, they both put in applications to operate daily services to Dulles Airport, Washington.

In order that these applications should be given fair consideration, it was decided by Mr Coleman, the Secretary of Transportation in the United States, to hold hearings in Washington on 5 January 1976 at which all those parties involved in the issue of the environmental impact of Concorde on the United States would have an opportunity to state their case. I was over in Washington when the hearings took place and, like everyone else, I was extremely impressed by the impartial way in which Mr Coleman conducted them. Those who were pro-Concorde and those who were anti-Concorde were at least agreed on the fact that Mr Coleman treated them all with equal fairness. He announced that within thirty days after the hearings he would pronounce judgement and, on 4 February 1976, he issued what will probably turn out to be a historic decision: that is, to permit, under certain conditions, the operation of Concorde into JFK and into Dulles for what amounts to a sixteen-month trial period.

As far as JFK is concerned, there are further complications brought about by the fact that the airport is controlled by the Port of New York Authority, that is to say essentially by the States of New York and New Jersey. In the case of Dulles Airport, outside Washington, however, this is a federally operated airport. Thus, although we can expect the controversy to continue to develop in the United States during 1976, it is likely to develop in a slightly different fashion in Washington than in New York. We can expect Congress to debate the issue. We can expect it to be debated in New York and New Jersey. We can expect action in the courts and it may be some little time before the situation is sufficiently clarified for Concorde services to commence operations into New York and Washington.

It is perhaps interesting in the New York situation to appreciate that, although a number of political and other figures have come out very strongly against the operation of Concorde at JFK, there have been people carrying a good deal of important opinion behind them who have been pro-Concorde. I recall that on the evening Mr Coleman made public his Decision to allow Concorde into the United States, my old friend Douglas Bader called me to tell me of a letter which had been sent to Mr Coleman by the New York City Central Labor Council of the AFL–CIO, of which he subsequently sent me a copy. This was a letter I hadn't seen before but which came out very strongly in favour of the operation of Concorde into JFK. This letter, signed by the senior officials of the New York City Central Labor Council of the AFL–CIO, concluded with the words:

> We are confident that the technical problems associated with the use of supersonic aircraft will be overcome. Minor inconvenience, though, is hardly reason enough to shut the gateways of New York City to progress. It is not only the clock that cannot be turned back; New York City cannot afford to turn back the multitudes of visitors who will profit by supersonic transportation.
>
> We therefore urge you to permit the landing of supersonic aircraft in our airports.

So we have a position now where Concorde is in operation, where the United States Government has taken a decision to allow Concorde to operate to airports on the east coast of the United States, and where that decision has apparently been endorsed by the President of the United States.

The next twelve months will show us where else Concorde services are to be developed, and where other routes can be opened around the world. They will also demonstrate the market appeal of the aircraft to airline passengers. They will therefore be absolutely critical months, not only to the Concorde project, not only to the works at Bristol and Toulouse, but to the European aircraft industry as a whole and to the future of high technology projects in Europe.

Twelve months from now, Concorde will either have transformed the pattern of airline travel throughout the world, leading to a new era as significant as that of the introduction of the jet airplane when it superseded the propeller airplane, or it will turn out to have been an interesting but expensive research and development project in Europe with very little in the way of practical commercial return.

I still believe very much that Concorde will turn out to have been one of the major developments in the commercial airplane business. I am confident of that.

This book is just my view, that is to say one man's view, of a project which started twenty years ago in the minds and on the drawing boards of people at Bristol. The book endeavours to trace the progress of that project from its beginnings to the mid-Seventies when it is an airplane in airline service carrying fare-paying passengers to their destinations.

I start from the premise that air travel is in essence about speed and that faster communication is to the benefit of mankind as a whole. I therefore make no pretence that this is a book by an unbiased and dispassionate observer. It is a book written by someone wholly committed to the human advantages of air travel. I have endeavoured to examine the effects on the Concorde project over the years of domestic politics, both in England and in France, of international politics, in particular in the United States, or rather as between Europe and the United States. I have tried to describe the way in which the technical problems of supersonic commercial aircraft have been resolved and I have examined some of the aspects of the environmental case against Concorde.

I come to the same conclusion today as was my belief at the beginning of the project, that supersonic travel is inevitable and that it will bring great advantages to the human race. I believe that the remaining problems are soluble and that in the future supersonic travel will be regarded as a perfectly normal way of moving from country to country around the globe. Supersonic transports will be developed which will be at least as environmentally acceptable as the best of the jet airplanes being built today. Indeed all the developments and improvements that we have seen in the past with subsonic jet aircraft will inevitably also come about with supersonic aircraft.

Geoffrey Knight
London, 1976

Concorde: The Inside Story

1 The Empire Syndrome

Those who first conceived Concorde were themselves brought up in a very different world from today's, when transatlantic travel in one hop was not considered feasible until the 1950s and aircraft were propeller-driven. Now we can get to the moon, air travel has shrunk the world to a series of overnight jaunts, and Concorde itself heralds an era of London to New York and back in the day with several hours to transact business between flights. Concorde was therefore planned at a time when the whole concept of world communications was undergoing radical reappraisal. Not surprisingly, it has been greatly affected by this time of confusion and change.

In order to understand the thinking behind Concorde and how it came to be designed as it was, it is necessary to go back for a moment in time and look at the influences that were brought to bear on the aviation industry and its planners immediately before it appeared.

Before the war, airlines, and even individuals, commissioned the design and construction of an aircraft for a particular purpose. It was normal practice for the Royal Air Force, when it wanted a new airplane, to ask industry to submit various proposals. From these it would select two or more to be built and flown in competition at their experimental station. This did not cost very much because aircraft in those days could be designed and produced for about a pound sterling for a pound weight, which meant that for £10,000 you could buy a new major aircraft. Quite often a firm whose design had been rejected might go ahead on its own as a speculative venture and still achieve success.

It would be fair to say that not all officially sponsored aircraft between the wars turned out to be good airplanes: the Battle, the Botha, the Defiant, the Whirlwind, the Manchester and others are all names long forgotten. Often those that did turn out to be a success had been privately sponsored. The Spitfire was developed directly from contenders for the Schneider Trophy contest. The Mosquito was the eventual development of a private venture entrant produced by de Havillands for the 1934 England–Australia race. The Blenheim stemmed directly from Lord Rothermere's imaginative sponsorship of 'Britain First', a newspaper aircraft to take his photographers to places in Europe and bring their pictures back in time for the chosen edition. The Manchester was so bad that it had to be completely redesigned with four engines instead of two, making it into an outstanding aircraft which they then called the Lancaster.

The Bristol Blenheim, when it was first produced in 1936, had a top speed of 250 miles an hour, which was fifty miles an hour faster than the single-seater fighters then in service with the RAF. That prototype was produced on contract, with half the money down on signature and the other half payable if the airplane was in the air twelve months later. The contract was only just honoured, because there had been a delay of three weeks in signing the contract and delivery would otherwise have been twenty days late. The entire cost was £18,750, which would hardly fit out the galley in a modern airliner.

Even just after the war many of the same rules applied. With commercial communications disrupted all over the world, it was felt that there was a demand for a rough air lorry, cheap and hastily produced, to be able to carry people and goods about. Four prototype Bristol Freighters were designed and produced for a total cost of £200,000. They weighed 37,500 lb and were sold for £37,500 each, based on the pound a pound price formula which had been the standard since 1920.

The war changed the aircraft industry, particularly on the

European side of the Atlantic. The Germans produced some remarkably advanced aircraft during the war, but defeat brought the liquidation of their industry and put them back a couple of decades. Both the British and the French had been inventive pioneers between the wars, but then the French industry barely survived German occupation. In Britain we operated at full stretch, but a clear division soon emerged between our output and that of our American allies when they entered the war. Britain concentrated on fighters and night bombers and the Americans on day bombers and transport airplanes. Night bombers and fighters had a short life expectancy. I am told that it was calculated at one time that the average life of a combat airplane was a mere fifty-three hours. Military air transports had to provide several thousand hours of service, and the Americans emerged from the war with this tremendous advantage of specialized experience.

Before the war ended, in fact in 1942, a committee had been set up under Lord Brabazon to consider what Britain would need in the way of civil aircraft after the war. Transatlantic air travel was not considered important; instead emphasis was placed on travel between the principal European capitals and on the old Empire routes which had been covered by Imperial Airways with its flying boats. Among the recommendations of the Brabazon Committee were two specifications for what were called Medium Range Empire and Long Range Empire transports. The longer-range version was intended to cater for such journeys as London to Cairo and Cairo to Karachi, which lie within the 2500-miles-range band. Travel across the Atlantic was still to be from London to Shannon or Prestwick, on to Iceland and from there to Gander and so to New York. At that time even the Americans, with all their experience, were catering for transcontinental flights with stops between New York and Los Angeles.

The Brabazon Committee's recommendations for these two Empire transports were put out to tender among the sixteen or so aircraft manufacturing firms which still existed then, and both of them were won by the Bristol Aeroplane Company, as it was then. They produced two aircraft which would have been real workhorses if technology in the form of the turbo-jet had not overtaken them. The first was named the Brabazon: only one prototype was built, which flew four hundred hours. The price of the whole operation was £6·45m. The second was the Britannia, which was the logical successor to the Empire flying boats. It was originally designed to be a thirty-two first-class or thirty-six tourist passenger plane weighing 100,000 lb, but it was developed and expanded until it ended up as a ninety-six-seater weighing 185,000 lb. We had bad luck with the Britannia: an engine caught fire in the second prototype, which caused many months' delay in bringing it into service, by which time the jet airplane was breathing down its neck.

During the early 1950s Britain concentrated on its highly promising turbo-props, and although delays and misfortunes upset the development of the Brabazon and the Britannia, Britain had a huge success with her medium-range Vickers Viscount. Initially designed for service within Europe, it was soon adopted by Australian, Canadian and American markets. 147 were sold in North America and final sales throughout the world totalled 438 to more than seventy operators.

But while Britain was concentrating on its turbo-props, the Americans were going for pure-jet power, beginning with the Boeing 707 and McDonnell Douglas DC8, which gained for the Americans a huge market advantage throughout the world which they have sustained ever since. Firms like Boeing, Douglas and Lockheed grew much larger than their British counterparts and Britain had to contend with a much smaller share of the civil and military market which in turn was being shared out among a substantial number of small- to medium-sized companies. The only British jetliner well advanced at the time, apart from the ill-fated Comet, was the Vickers-Armstrong V1000, which was initially a military transport, but it was cancelled by the British government because of lack of interest in its commercial version by BOAC.

One other thread to be remembered in the Concorde story is the recovery of the French in the 1950s from the desolation of war. Two major companies had been built up out of nothing at all. One of them was Dassault, which concentrated almost exclusively on military aircraft. The other was Sud Aviation in Toulouse where, under the leadership of a remarkable and determined man named Georges Héreil, they had succeeded in producing a highly acceptable twin-engined short-haul jet called the Caravelle. Héreil managed to sell a batch of twenty of them to America. In a remarkably short time he had re-established the French industry in the international airliner market.

The Caravelle embodied a substantial quantity of British equipment. It had Rolls-Royce Avon engines, a Dowty undercarriage and a good many other British components. The French considered that because they had supported British industry by buying such a substantial quantity of equipment, which formed quite a high proportion of the price of the airplane, British European Airways should purchase it for their own operations. Almost every other major European airline did so – Swissair, SAS, Iberia, Sabena and Alitalia among them. Instead, BEA commissioned Vickers to develop the Vanguard for them. The French, although not enough people in Britain took account of it, regarded this as tantamount to treachery on our part. When the time came for collaboration on Concorde, this undercurrent of feeling was there and was to cause its problems.

By the mid-Fifties at the Bristol Aeroplane Company three new airline projects were being devised. The first was for a high capacity short-haul, three-engined commercial aircraft, the Bristol 200, which could have been the natural successor to the Vickers Viscount and the French Caravelle. The second was a NATO military project, the Bristol 208/224. The NATO technical directorate was making a serious endeavour to write specifications for its military requirements which would avoid the multiplicity of different types with which the alliance would be encumbered in the event of a conflict. The situation was a mess and is still not much better, with different types of fighter, of guided weapon, even of ammunition. One requirement was for a short take-off military transport, a combined troop-carrier and supply aircraft to operate somewhere up near the front. Bristol put forward a design in which we used four Harrier-type Rolls-Royce Pegasus engines with the jets deflecting a substantial amount of thrust downwards to provide lift for vertical take-off. We had even studied joint production with Canada, as the Canadians had built a version of the Britannia, the Canadair CL28 Argus, under licence as a military anti-submarine airplane. If Bristol won the competition, they would take Canada in as partners, and if Canada won it with another design Bristol would collaborate with them. It all looked very promising for a time, but then everything collapsed. The third airplane on which we had done a lot of serious work was a supersonic transport, the Bristol 198, although our thinking did not then assign it the immediacy of the other two.

At that time our chief designer, Doctor, now Sir, Archibald Russell, known to everyone as 'Russ', was a member of the Air Warfare Committee. They had come to the conclusion that if Britain were attacked by subsonic jet bombers we might have a chance of shooting them down on the way out but no chance whatsoever of shooting them down on the way in. Guided weapons provided part of the answer, and at Bristol we also evolved a most successful surface-to-air missile system called 'Bloodhound', which continues in service today. However, after a lot of analysis and consideration it was decided that a Mach 3 fighter – flying at three times the speed of sound – might have a chance with a few minutes to spare, providing it had the right radar guidance for intercepting the bombers on the way in and shooting them down before they got to London instead of after they had set London alight.

This resulted in Bristol getting a Ministry of Defence contract to build two Mach 3 research airplanes. They were called Bristol 188s, and made of steel. They were absolutely seminal aircraft in the Concorde story.

The Bristol 188 concept marked a triple watershed. First, it provided the last example of the faster defence fighter.

Two of them were developed and flown at an overall programme cost of £20·5m. However, the manufacturing processes that went into this machine were so difficult and expensive to construct that Archibald Russell ruled out steel as a raw material for supersonic application, and this experiment was later to become a most significant factor in the choice of aluminium alloys for the supersonic civil airplane. He also made up his mind at the same time that the civil aircraft would fly at Mach 2, twice the speed of sound, this being the upper limit for the use of known aluminium alloys at the exceptionally high temperatures that would result from the kinetic heating phenomenon – an opinion to which he adhered through thick and thin in the years that lay ahead in spite of many pressures to the contrary.

However, the apple of Russ's eye at the time was his three-engined medium-haul jet, which he was convinced could be a world-beater. We were unable to convince BEA that this was what they needed and they eventually settled for a smaller aircraft of lighter weight, the de Havilland Trident. With hindsight, one reason for this may well have been that the Bristol project was naturally based on a Bristol engine whereas BEA was by this time strongly wedded to Rolls-Royce in view of their great success with their Dart prop-jet in the Viscount.

In an attempt to find an outlet for his design, Russ and several others at Bristol, including myself, tried to interest Pan American, who put the specifications through their comprehensive analysis system and, after a considerable redesign and readjustment, looked like settling on a version with a weight of 150,000 lb compared with the 100,000 lb of the Trident. In the end nothing came of these negotiations.

There is one final element in the build-up to Concorde that needs to be recorded. Russ's three-engined airliner was to have been fitted with the Bristol Olympus jet engine. This had been developed originally for the Vulcan bombers, and the man who came in on the initial design stage, and has stayed with it ever since, was Sir Stanley Hooker. In the Fifties the Olympus jet engine had a thrust of 10,000 lb, remarkable for an engine of its weight. It was rapidly developed up to a thrust of nearly 20,000 lb, and would have had to be derated to 13,500 lb for the Bristol tri-jet project. It was also the engine foreseen for the ill-fated TSR2 military project. It was the one British engine capable of being adapted to sustained commercial performance at twice the speed of sound.

This then was the scene which led to the thinking that produced Concorde: initial misconceptions after the war about the necessary range and capacity of new aircraft caused by thinking still steeped in the old Empire routes; delays, disaster and cancellation in the British attempts to realize aircraft to span the Atlantic; too much concentration on shorter-range regional models; the massive predominance of the American aircraft industry already aiming at a near-monopoly of the long-haul market; the end of the era when it was possible for individual aircraft firms in Britain to conduct research and development work at their own expense at reasonable cost, even with modest initial government support; the re-emergence of the French as powerful rivals in the European aircraft production field.

On the other hand, we still had considerable cards to play: continuing experience in aircraft production exceeded only by the Americans; designers of genius; considerable experience of tested supersonic flight; research and testing establishments; and one of the most advanced jet engines in the world with the Olympus. The scene was set for the great leap forward that would take us out of the ruck of mediocrity and put us years ahead of our competitors.

2 Leap Through the Sound Barrier

The latter part of the Fifties was in retrospect a critical period for the British aircraft industry. In the decade since the end of the war we had spent a good deal of money, exhibited a great deal of activity and ingenuity, and produced a series of turbo-prop and jet commercial aircraft, none of which, with the important exception of the Viscount, had turned out to be successes in the airline business. The Comet had failed to provide the hoped-for technological breakthrough because of the disasters that overcame it in the early 1950s, although the meticulous care that was taken to remedy its defects still made it, in a way, the herald of the jet aircraft age. The Britannia had been overtaken by the American jets and was probably obsolescent as soon as it flew.

The Americans had three factors overwhelmingly in their favour and did not fail to take advantage of them. First they had the experience of a world market which derived from their position as the suppliers of transport aircraft to the Allied forces during the war, so that they were able, as soon as it was over, to assume a position as suppliers of civil transport aircraft to most of the Western world. At that time, of course, these were still piston-engined airplanes. The Americans had greater manufacturing resources and far and away the biggest domestic market in the world. They also recognized at an early stage that the jet engine would be the propulsion unit to supersede all others in the commercial airplane business, and they put massive resources behind its development. By the end of the Fifties and in the early Sixties the Boeing 707 and the

McDonnell Douglas DC8 were poised to take over the whole of the long-haul commercial airplane business and virtually did just that.

At that time there was still doubt in most people's minds as to whether it would be possible to develop an economically viable short/medium-haul jet engine airplane, and in the late 1950s this still had to be proved. The French had made a breakthrough with their Caravelle, but its economics were never really its strong point. It carved out a reasonably substantial place for itself in the market because airline customers preferred to fly in a jet if they could rather than behind a propeller, and its loads were high. It certainly did not have the kind of economics that later developments in the short/medium-haul jet field like the Boeing 727 and 737, the McDonnell Douglas DC9 and the BAC One-Eleven had. In Britain we did develop the de Havilland Trident, which was the airplane eventually selected by BEA for its short/medium-haul routes, but it never really managed to find itself a foothold in the world market because the Boeing 727, which came on at a later stage, proved beyond question the most massively best-selling airplane that the commercial airline market has ever seen. The old wartime DC3 Dakota may just have had the edge in numbers sold but then it also had a military market during World War Two.

Britain still had a tremendous reserve of will and skill available. We had too many companies, too few outlets, and government funds had to be spread too thin to keep them all going, but there were plenty of people aware enough of the dangers that lay ahead to take positive action. The time had come for the government, the companies, the designers and specialists to get together and chart a course for the future.

We have in Britain a remarkable organization, the Royal Aircraft Establishment, based at Farnborough. Its director in those days was George Gardner. From their comprehensive knowledge of the current climate of aeronautical research, they had started applying themselves to the whole problem of the next generation of airliners. By early 1956 they had come to the conclusion in theory that the inevitable next stage must be the development of a supersonic commercial transport.

It was time for government to intervene. The aircraft business had already reached such a degree of sophistication and international importance that it had become impossible for any individual firm to find the solutions which would guarantee employment in the industry and ensure for Britain the place amongst the leaders in the field which it had so long occupied.

George Gardner presented his conclusions to the government, which reacted for once with speed and consequence. They set up a joint government–industry committee, including BOAC and BEA, to inquire into the whole field of activity and recommend necessary future action. It was called the Supersonic Transport Aircraft Committee and held its first meeting on 5 November 1956, under the chairmanship of a senior government official named Morien Morgan, whose current job as deputy director of the RAE was to coordinate the activities of the various government research establishments. Its membership was largely technical, and it was probably one of the most high-powered and effective committees ever formed for a specific major purpose. All the major airframe companies were represented – Avro, Armstrong Whitworth, de Havilland, English Electric, Fairey, Handley Page, Short Brothers and Harland, Vickers-Armstrongs and, of course, Bristol Aircraft with Archibald Russell as its representative. The engine makers were there too – Armstrong-Siddeley, Bristol Aero-Engines, de Havilland and Rolls-Royce. The airlines, the ministries and the research establishments contributed their best men.

The Committee's purpose was to discover whether it was possible to produce a commercial aircraft capable of flying at supersonic speeds while maintaining established standards of passenger comfort and safety and meeting the airlines' normal requirements of profitable operation. Supersonic flight had been an established fact for some time, but it had only been applied to military aircraft, where the operational requirements are quite different and commercial economic considerations do not obtain. A supersonic fighter only has to operate for a relatively short period of time, and can be given engines of enormous power which would be hopelessly uneconomic for commercial purposes.

In technical matters, I rely on my old friend Archibald Russell, to whom I am indebted for the following exposition. He tells me that the solution to the problems of supersonic flight lies essentially in finding the correct lift/drag ratio. The faster a plane flies, depending on its shape, the higher its drag (the downwind resistive force on the aircraft resulting from its shape and the airflow pattern around it) as it has to force its way through the atmosphere, and the greater the thrust the engine needs to push it along. This affects the range rather than the speed of an aircraft. You can always pile on enough power to push a fighter through the sound barrier for a few minutes. In the early days they used rocket boosters for the purpose but this is not a solution that lends itself to commercial aircraft. These have to travel long distances on scheduled flights and overcome air resistance, which raises the temperature of the metal frame of the aircraft, and all without inconveniencing passengers or requiring so much fuel that the aircraft would never reach its destination.

STAC, as the Committee was called from its initial letters, set out to study these problems. They were given a very loose brief and spent £700,000. They were at liberty to study all the possibilities between the speed of Mach 1·2, just over the speed of sound, right through to Mach 2·7. It was not laid down whether the airplane had to go a thousand, two thousand or three thousand miles. Nor was it specified whether the plane would carry ten or a hundred passengers. The first task was pure research to find out how to improve the lift/drag ratio from the maximum that had been recorded on military aircraft, which flew supersonically by brute force, and adapt the findings to a civil airplane that had to fly a long way. The overall problem was to find an acceptable total aircraft solution, and a number of technical subcommittees were set up – a power plant committee, a structural committee, a systems committee, a navigation committee – each required to make its contribution to this common end.

The key figure in all this was Dietrich Küchemann, a German who had come over to England shortly after the war and was one of the leading aerodynamic theorists of the Royal Aircraft Establishment. He had produced a new principle of supersonic flight which went against all the precepts of aerodynamics over the previous sixty years. Aircraft designers had been trying all that time to devise aircraft shapes which would eliminate or minimize what is called vortex flow – the pattern of air turbulence caused by an aircraft forcing its way through the atmosphere – which was regarded as very extravagant and energy-consuming.

It had long been supposed in the older and more simple days that the best way of coping with this was to have as straight and slim a wing as possible. Once you started sweeping the wing back it made higher speeds possible but the wing became very susceptible to disturbances in the atmosphere which tended to make the plane unstable. In a side gust, the effect of the wind on the wing on that side is entirely different from the effect on the wing on the other side, and the more you sweep the wing back the more severe that problem becomes. Already the civil subsonic airplane wings had what are called 'fences' on the leading edges, which made it possible to reduce this sort of turbulence and keep the flow of air smooth over the wings. When a wing stalls (due to the breakdown of the airflow over it and hence the loss of aerodynamic lift) it must stall from the root (fuselage side) first. If it starts to stall at the wingtip, the airplane quickly falls out of control.

Dietrich Küchemann had a revolutionary solution to this problem for a supersonic airplane. He wanted an arrow-shaped wing which created such large vortices (rolls of air curling over the sharp leading edge) that substantial extra natural lift resulted. Moreover, because these vortices were so strong that they could not be influenced by the kind of airflow disturbance which would upset a conventional wing dependent entirely upon attached airflow, additional aircraft stability was also provided. All this is worth recording in layman's terms because it provided the basic breakthrough in aerodynamic thinking which has made supersonic flight possible and safe.

There is another factor, the sonic boom, which is a basic characteristic of supersonic flight and which must be understood before the problem it creates can be overcome. When an aircraft is flying at less than the speed of sound, the pressure distribution round the plane sends out pressure signals in all directions and it is able to send those signals forward to warn the air ahead: 'Look out, there's an airplane coming, you'd better get out of the way.' These pressure signals travel in all directions at the speed of sound. The air then has time to flow in streamlined fashion around the body and the wing of the aircraft. When you fly supersonically, you produce a bow wave from the nose of the plane exactly as you do in a ship. The air is no longer streamlined. The airplane gets there before the signal and it meets the unwarned air with a bang.

The other absolutely basic principle that emerges from this is that the shape of an aircraft designed for supersonic flight must bring the entire structure within the bow wave that it causes. If any part of the airplane projects outside, then it sets up a subsidiary bow wave. That is the essence and the explanation of the delta wing shape. The compromise that had to be found was for a delta wing which would be as efficient as possible flying at a given supersonic speed but would still provide normal lift, stability and control at subsonic speeds when landing and taking off.

Every other problem was subsidiary to this – the power of the engines, the necessity of dispersing the heat generated in the metal skin of the aircraft by its high speed, the differences in trim and balance of the aircraft at different heights and different speeds, and how to fit all the controls, fuel tanks, hydraulic gear – and passengers – into the optimum shape. There is one further technical problem that had to be overcome – here again I am indebted to Archibald Russell – and that is that the temperature of the aircraft skin increases as the square of the speed in hundreds of miles an hour. If that sounds too technical, it is really quite simple. If an airplane is travelling at 100 mph it increases the temperature of the air hitting it by 1°. One squared is one. If it is flying at 500 mph, the skin temperature increases by 5 squared: 25°C. However, as even subsonic jet airplanes fly at a considerable height, where the air temperature is probably −50°C, the skin temperature is still 25° below freezing at 500 mph. If you are flying at 1000 mph, the temperature increases by 10 squared, that is to say, up to 100° minus the 50° of the cold high air. At Mach 2, which is around 1350 mph, 13·5 squared is 182° and even allowing for the −50° for the cold air, that leaves you with around 130° as the temperature of the airplane structure. This is the temperature which heats the fuel, the control surfaces, the lubrication, the electric cables and every component outside the pressure cabin.

This is one of the reasons why Archibald Russell became such a constant protagonist of Mach 2 as the correct speed for supersonic flight. At Mach 3, which is around 1700 miles an hour, the skin temperature is 300°, and applying the −50° still gives a net 250°C. None of the known fuels would have stood it without boiling, the lubricating oils would not have operated and plastic windscreens and windows would have given serious trouble. None of the electrical installations known at the time would have worked.

Considering the technical problems that had to be overcome, involving endless testing of metals and components, the Supersonic Transport Aircraft Committee worked remarkably quickly. It produced a considered report on 9 March 1959. The main conclusion was that supersonic air transport was economically feasible. Two theoretical alternatives were proposed: a medium-range (1500 n.m.) aircraft cruising at Mach 1·2 and a long-range (3000 n.m.) aircraft cruising at Mach 1·8. Development costs were estimated at £50–£80m for the smaller aircraft and £75–£95m for the larger. It was estimated that the planes would cost about £3–£4m each to buy and about one and a half times as much to operate as subsonic planes.

The Ministry of Defence, with Duncan Sandys as its head

at the time, reacted swiftly. In September 1959 it commissioned feasibility studies from Bristol for an SST (supersonic transport) with a slender wing and a conventional discrete fuselage, and from Hawker Siddeley for a thinner arrow-shaped flying wing with an integral passenger compartment. The airframe materials were to be principally aluminium alloys, using one which had already proved itself in the pistons of conventional aero-engines. The use of steel or titanium was ruled out because of the difficulties that had arisen in the construction of experimental military aircraft.

It was now up to the aircraft companies to respond to the momentum that events were acquiring. It has to be said that Bristol had got the better share of the feasibility study deal. Although the flying wing seemed likely to have the best aerodynamic efficiency of any shape, this was more than offset by the structural weight penalty it would carry. Archibald Russell had long since settled in his own mind for the discrete fuselage with a delta wing, although consideration had been given by Morien Morgan to other shapes such as a gull wing, what is called a canard, with a little wing at the front, and even M- and W-shaped wings. The trump card was our possession of the Olympus engine. At the time it was still intended for use with the military TSR2, and this meant that a good proportion of the development costs would be met by the defence budget.

In the event, both firms delivered their studies six months later, in March 1960. Russ had taken a hard look at the specifications and had come up with an enlarged Mach 2·2 airplane weighing 385,000 lb and carrying 132 passengers across the Atlantic. This was going to mount six Olympus engines in a bank on the top of the wings at the back. The Ministry was horrified by the size and weight, especially as the Boeing 707 was taking shape with an all-up weight of only 300,000 lb. Nevertheless the Hawker Siddeley version fell by the wayside and Bristol was given the go-ahead for the next stage of a design study, which was presented in the following August and carried the designation of the Bristol 198.

The six engines were not, in fact, a good idea as access for maintenance to the middle one in each set was extremely bad. To meet the objections that arose, we carried out a modification for a four-engined, hundred-passenger version which was delivered in late 1961 and was known as the BAC223. We had got the weight down to 270,000 lb, which was much more likely to gain support for development from the Ministry. As a design it was the immediate forerunner of the Concorde, although there was a long way to go yet. The government had made up its mind that aircraft development on this scale was beyond the scope of the economy to finance on its own. Both Hawker Siddeley and Bristol had been instructed, as a condition of carrying out the feasibility studies, to explore the possibility of collaboration with American, French or German manufacturers. International competition was about to give way to international collaboration.

I cannot leave matters at this point without paying a tribute to the chairman of STAC, now Sir Morien Morgan, and Master of Downing College, Cambridge. This lively, voluble Welshman had been the driving force on his Committee, with more influence than everyone else put together. He kept his team of the country's leading scientists working in happy unison and is entitled to great credit for a quite remarkable example of technological achievement. It was a mark of the advanced state of British aircraft technology at the time that, although the supersonic transport was a great leap forward, it was based substantially on known and tested technology. The airframe designers probably had the easier part of it. The structures and systems specialists had to push their knowledge well beyond the limits of current practice because the obligation of getting everything to work in an envelope substantially above the boiling point of water was a completely new problem which they solved triumphantly. In the long-drawn-out years that lay ahead, hardly any of their calculations were fundamentally wrong. By all accounts it was an astounding performance.

* * *

While all this was going on, the British aircraft industry was faced by the biggest upheaval in its history. Throughout this Concorde story, it is necessary to refer from time to time to the effect that changes of government have had on the project. What gets overlooked is that aircraft development takes a longer span of time than the constitutional life of governments. Even in the less complicated period that we are still dealing with, major commercial aircraft still took four to six years to emerge from the drawing board to commercial operation. Unless re-elected, no government can last more than five years at most and, with public money playing an increasingly essential role in the financing of major projects, this has meant at least one major reappraisal by the Treasury during the production process of any major aircraft and, in the case of Concorde, several.

In November 1959 the Conservative government was re-elected for the second time since it first came to power in 1951. Much had happened during that period. American military aid had been sharply reduced in the latter 1950s and Britain had to revise her military strategy and expenditure to match the capacity of the economy. The Defence White Paper of 1957 had already radically altered the outlook of the British aircraft industry as far as future lucrative defence contracts were concerned. With its renewed mandate at the end of 1959, the government had decided to do something about the excessively fragmented aircraft industry it could no longer hope to sustain.

Under extreme pressure from the responsible Minister, Duncan Sandys, the industry was massively restructured. Of the six previous engine manufacturers, only two emerged. Rolls-Royce remained intact and Bristol-Siddeley was formed out of the remainder. Only two major airframe groups survived. One was Hawker Siddeley, which now took over de Havilland, Blackburns and Follands, and the other was the British Aircraft Corporation, formed out of Vickers-Armstrongs, English Electric and Bristol Aircraft. Bristol had always been one of the most fiercely independent firms in the country, but when it was made clear to them that such a grouping would be the only hope of sharing in future major contracts, they had no choice but to conform. Bristol was already involved in work on the supersonic commercial transport and was not going to let that slip out of its hands as well.

Duncan Sandys held many cards in his hand when it came to forcing people to do things that they might normally find distinctly unpalatable. He was suggesting: 'Unless you do this and put yourself in a position where you have the sort of capital and strength to compete with America and the rest of Europe, then I'm not going to give you any financial assistance from my department.' Moreover anyone who ever had to deal with Duncan Sandys knows that once he got an idea in his head he was very nearly unstoppable. He certainly managed the merger operation in an amazingly short time and, looking back on it, most of the participants are probably still shaking their heads and wondering how he managed to push it through. I rather doubt whether any of them particularly regret what was done, in the long run, and I consider that he did the industry a favour.

The intention when the BAC merger took place was that the proud old names should be continued in view of their influence throughout the world, and that there would be three subsidiary companies of BAC to be known as Bristol Aircraft Limited, Vickers-Armstrongs (Aircraft) Ltd, and English Electric Aviation Ltd. For a time, this is what happened.

What Bristol, Vickers and English Electric did was to put their airframe interests into one pot and take stock in the new company in exchange for the assets they were putting in. Bristol got the thin edge of the share-out, with twenty per cent of the shares in BAC, Vickers had forty per cent and English Electric got forty per cent. The three main component companies all maintained their boards and executive directors and a new board, mainly of non-executive directors, was set up for the British Aircraft Corporation. However, this did contain as exceptions one very strong character, Sir George Edwards, who came from Vickers as the Executive Director for Aircraft, and Lord Caldecote, who came as Executive Director for Guided Weapons.

The whole arrangement officially came into effect on New Year's Day 1960. In practice it took most of 1960 to put BAC together properly as a company. The new Corporation took over the supersonic air transport design project on which Bristol had spent so much time and thought, and we all braced ourselves for the next stage – collaboration with the French on this breakthrough in the air transport field.

3 Anglo-French Marriage

The French provide half the Concorde story, and, as might be expected, the history of Anglo-French collaboration on the project has not always been entirely smooth. Differences of national approach, differences in organization, separate traditions of technological experience and a high degree of national pride on both sides of the Channel have all contributed to the difficulties that arose from time to time in what was, after all, a completely new experience. There was no precedent for the type of joint undertaking that presented itself at the end of the Fifties and the beginning of the Sixties in a major international project of this nature involving two languages, two different forms of government control, two differently organized industries and two completely different methods of approach.

I need to digress for a moment to look at the situation in France. We have seen how the British aircraft industry continued into the late Fifties on very much the same individual, fiercely competitive basis which had characterized the first fifty years of its growth. The French had also been worthy pioneers in aviation. Both before World War One and between the two world wars they could claim equality in many respects with their British counterparts. Then, in 1936, they underwent a change in structure which has only become a possible factor in British industry today, forty years later. Under the Popular Front government in power at the time, the French aircraft industry was nationalized into six regional groups.

This form of organization persisted, although two of the dispossessed private firms, Breguet and Dassault, managed to maintain an existence outside these groupings. In the difficult years after the war, the state enterprises found themselves spurred on by the competition of groups they had ousted, which kept them very much on their toes. Their administrators had to be sharp businessmen as well, and an outstanding figure emerged in the person of Georges Héreil, who had started his professional career as a lawyer and trustee in bankruptcy for the tribunal of the Seine. He became president of one of the reconstructed nationalized firms which had the laborious French title of 'Société Nationale de Constructions Aéronautiques du Sud-Est', which was usually abbreviated to SNCASE.

We do not need to dwell on its history for long. Its first commercial transport was based on a pre-war design. The French industry, even in those hard times, did have an extraordinarily gifted hard core of trained aeronautical engineers who were searching desperately for an outlet for their talents. They were full of ideas. They had to develop aircraft quickly at minimum cost and get the prototypes into the air in a minimum of time. They also helped other firms to remain in existence by spreading the subcontracts amongst themselves. In 1949, the similarly-named SNCASO, with the SO standing for Sud-Ouest, obtained government funds to develop an airliner with four jets not dissimilar to the British Comet. This was beyond the capacity of government funds and the industry at the time, and it was decided that the most economocal project would be to develop a medium-range twin jet. Georges Héreil's Sud-Est company submitted a proposal for this and won the contract. It became known as the Caravelle, which went into service in 1957 and enjoyed enough success to enable Héreil to absorb the Sud-Ouest group and form Sud Aviation. They were the people with whom we were called upon to deal.

Héreil's engineering and design team, headed by two people who will play a substantial part in this story, Pierre Satre and Lucien Servanty, were already concentrating on the next stage. They had decided that they would not challenge the Americans head-on, but would concentrate on developing a supersonic airliner to serve the same sort of market, but faster than the Caravelle. Although their technology was probably not as advanced as that of the British at that time, the French had already produced several supersonic research and fighter prototypes and were catching up fast.

Meanwhile we had been trying to find some accommodation with the Americans. Sir George Edwards had been to the United States several times and spoken to the leading American aircraft manufacturers to see if there was any basis for collaboration. The stumbling block with the Americans was that they were simply not interested in any leap forward which did not involve a speed of at least Mach 3 – three times the speed of sound. BAC Bristol, under the prompting and guidance of Archibald Russell, had made up its mind that Mach 2 was the only feasible speed that lay within the limits of known technology and operational requirements. He was strongly supported by Sir George Edwards, who had by now become very much the dean of aviation in Great Britain. We still had something of an inferiority complex as far as the Americans were concerned, as their development of long-range aircraft had outstripped us in the postwar years. We were worried that if we settled for a Mach 2 airliner, and the Americans with their vastly greater resources and speed of operation developed a Mach 3, then we were dead ducks. Our design would be obsolete before we'd punched in the first rivet.

So under prodding from Duncan Sandys and the Ministry, the British began conversations with their French counterparts. At first little real basis for collaboration was found. The trouble was that although Sir George Edwards and Georges Héreil knew each other perfectly well, they were both too strong and determined to find it easy to share a common approach. Each had an immense record of achievement behind him, and it would have been extremely difficult to get a combined project off the ground involving both of them. In any case Sir George was having to give a considerable amount of time to setting up the new corporate organization of BAC.

Then at the Paris Air Show of June 1961, Sud Aviation

displayed a model of a Super Caravelle designed to convince the world that the French were in a position to produce a four-engined, seventy-passenger supersonic plane with a range of up to 2000 miles. The pressure was on the British.

The Bristol team of Dr Archibald Russell and his Chief Engineer, Dr Bill Strang, began having conversations with their French counterparts Satre and Servanty. For some time they did not get beyond the stage of trying to find a use for common components in their separate projects. The Bristol people never really lowered their sights from the conception of a long-range transatlantic airliner, and the French never raised theirs from the basic idea of a medium-haul supersonic airliner for use on overland routes. So they talked mainly about using the same engine, the same hydraulic pumps, the same electric generators, the same air-conditioning system, the development of the powerplants in common, with the costs to be shared between the two companies and governments, and then each side would go its own way on the airframe. This parallel thinking continued for two or three years and was to a considerable extent exacerbated by a conviction on the British side that they were far ahead of the French in technology and that the newly formed British Aircraft Corporation carried far more financial and commercial clout. This was all very well, but the days of self-financing were over.

At this point the two governments intervened. In September 1961 the French Minister of Transport, Robert Buron, proposed a meeting with the then British Minister in charge of Aviation, Peter Thorneycroft, to resolve the deadlock. The result was a firm directive to both BAC and Sud Aviation to coordinate their designs and come up with a compromise airliner.

On the engine side, this caused remarkably little difficulty. Bristol-Siddeley held the whiphand with its already proven Olympus, and soon reached an agreement with the French engine firm of Société Nationale d'Etude et de Construction de Moteurs d'Aviation (SNECMA) to develop the powerplant further on a joint basis. They signed an agreement in

November of that year, and it was formalized and approved by both governments by March 1962.

On the airframe design side, things hung fire. Not only were the engineers and designers failing to find common ground, but they were also wrangling over who should be the dominating partner in any eventual collaborative project. BAC was not going to surrender its predominance and experience without a fight, and Georges Héreil, who had achieved extraordinary success in a very short time, was not going to surrender the ground he felt he had won. In November 1961, when the two firms presented their first conclusions, the situation was still confused. They recommended no less than four design projects, with each side putting forward its own version of aircraft to meet the long-range and medium-range requirements. Each had adapted its own main theme to the other, and none of the results was satisfactory by any objective criterion.

The governments intervened once more. Peter Thorneycroft and Robert Buron met again in December 1961 and instructed their respective industries to come up with a joint outline project. Still no real progress was made and the discussions and negotiations dragged on through the spring of 1962. I must not suggest that this was altogether a stand-up fight. Each side realized that the project was dependent upon government finance to get it off the ground, and several gestures were made at industrial level to improve the atmosphere. Sir George Edwards made a point of subcontracting a substantial element of the tail assembly of the Super VC10, which Vickers had brought in to BAC, to the Sud Aviation subsidiary factory at St Nazaire. It was a straight commercial transaction done deliberately to soften the French attitude.

At this point, too, political considerations intervened. De Gaulle had come to power in France, and the British government was in the process of conducting the first set of negotiations to enter the Common Market. Julian Amery, who took over from Peter Thorneycroft as Minister of Aviation in July 1962, has told me of the serious attention that the British government gave at the time to establishing European partnership at every possible level, and in particular Franco-British partnership on such projects as the supersonic air transport, chiefly as a means of asserting European independence of America. The government was anxious to establish the closest cooperation with the French, and indeed with the Germans if they could. Their appraisal was that the French aircraft industry was the only one that was within striking distance of ours. The Germans were still operating only in a very small way.

Julian Amery, who spent a great deal of his time in Paris during these months, negotiating with both their defence and civil sides on other joint projects, was conscious of the fact that British technology was still ahead. On the other hand, he was working on the basis that with the enormous sums involved in development costs, it was impossible for either government to go it alone. So the government was prepared to make the sacrifice of knowhow in return for the French making a comparable financial contribution. With the two firms still dragging their feet, the two governments intervened. Early in 1962 it was laid down that the chairmanship of the joint industry group studying the problem should alternate between the British and French representatives. At this point Georges Héreil resigned to take up the chairmanship of the Chrysler Motor Corporation's French subsidiary, Simca.

In his place, the French government appointed as president of Sud Aviation General André Puget, who had been Commander of the Free French bomber squadrons flying from Britain during the war and was a noted Anglophile. It was a counter-gesture with far-reaching results. From the time of his appointment, things started moving. That is not to say that the personal and professional problems of contact between strong-minded and gifted men on both sides were resolved for all time. Far from it. The rest of the Sud Aviation top team, led by Louis Giusta, together with Satre and Servanty, remained intact. On our side, Archibald Russell and Bill Strang retained their functions in full. The clash of strong wills and strongly held beliefs continued for years, but there was progress and it was immediate. Not the least important element was that Sir George Edwards and General Puget got on extremely well, and their mutual trust and collaboration stood the joint project in good stead through its crucial formative years.

Two myths have grown up about the final stages of the distinctly frenzied Anglo-French negotiations to develop the supersonic airliner jointly. One was that a copy of the report of the Supersonic Transport Aircraft Committee had been handed clandestinely by a British ministry official to the French. This has never been proved conclusively one way or the other. If it did happen, the French would have found little in it to surprise them. Technology makes parallel advances in most countries, and there would have been little in the technical conclusions which they would have found new or illuminating, although they certainly provided clear evidence of the firm grip the British had on the technical problems involved. Either way it made little difference, as the rival British long-range and French medium-range conceptions continued to cloud development for years.

The other story which has gained some public currency is that there was a secret meeting between Bill Strang from Bristol and Lucien Servanty, Sud Aviation's chief designer, that they went off to a small suburban office and, while everybody was looking for them and wondering where on earth they were, sketched out the first basic common design for the supersonic transport. The reality is much less romantic. In the autumn of 1962 the two men did indeed meet, with Archibald Russell and Pierre Satre present. In a session of furious activity they hammered out a common approach to both designs. Neither side was happy about such a rushed solution, but government pressure was on and they did produce a twenty-page report with a generalized drawing covering both the long- and medium-range versions to be submitted to their governments. All their assumptions were submitted to endless testing in wind tunnels and on test benches and in due course considerably refined, but at the time it served its purpose.

However, this hastily conceived document did contain within it major seeds of future differences and conflicts of opinion. Mick Wilde, now the British Project Director on Concorde, who was also present at the meeting, remembers that different interpretations were put on it by both sides in the industry. The British worked on the assumption that a design had to be tested in all its aspects if it was to become reality. The British government was accustomed to this process in its aircraft industry and so presumed that all theory would be tested before it was put into practice. The French worked differently. Their lean years after the war had taught them, so Mick Wilde thinks, to rely much more on hunch. They brought tremendous enthusiasm to the new project and, provided that a design appeared to have been expertly worked out, they tended to interpret this in terms of jigs and tools, and to cut metal and construct the airframe much faster than the British.

In spite of the flair of the French, however, Mick Wilde remembers them as being much more rigid in the interpretation of a given specification. Once the overall concept filtered down to the design and construction staff, they tended to regard it as an edict to which they had to conform in every detail. This discipline brought quick results, but it was quite foreign to British practice. We tend perhaps to go too much the other way. Almost everybody who works on a project acts as though he were the chief engineer and considers it his right and duty to question every assumption. The problems that arose from these two different national attitudes still lay in the future, but they were one of the causes of the delays and the expense that ensued.

Those of us, and that is most of us, who saw in the supersonic air transport project a means of leapfrogging American dominance in the long-haul field can derive vicarious satisfaction from the story that Julian Amery tells, with authority, of the final stages before the signature of the Agreement with the French. Not only were the Americans engaged in similar research, but they were extremely perturbed at progress that was suddenly being made on the other side of the Atlantic. The immediate ex-Chairman of the World Bank, Eugene Black, a man of great power and influence in the inner councils of the United States, arrived in London and asked to see Julian Amery. The Minister invited him to dine. The time was November 1962.

After preliminary talk about exchange rates, and the world economy generally, Eugene Black launched into a long attempt to persuade the Minister that the whole idea of a supersonic aircraft was wrong. It would have to be subsidized. It could not be a private enterprise operation and it would tie up substantial resources. He was clearly well briefed, and made a carefully considered and very lengthy attempt lasting right through the evening to dissuade the British government from going ahead with its plans. Julian Amery drew the natural conclusion that if the Americans were so keen that Britain should not go ahead with the project, the gamble was probably going to pay off. The Cabinet took a preliminary decision to go ahead.

Shortly after, Julian Amery was in America as the guest of Najeeb Halaby, a powerful figure in the industry, who was at that time Head of the Federal Aviation Administration. He was in charge of a project to study the building of a supersonic airliner on the American side. At their first lunch, Julian Amery asked him: 'How are things going?' Halaby told him that he was confident and that the President would appoint a very high-powered committee to supervise the operation. Amery asked who was on it, and Halaby produced a list from his pocket which named Eugene Black as chairman. Amery reported this back to his colleagues in London with the recommendation that he thought the British case was stronger than ever, as he had grave doubts as to whether the American aircraft, which was still being thought of in terms of Mach 3 operation, would ever get off the ground.

Julian Amery, who has been a strong proponent of Concorde ever since, derives great satisfaction from his attitude. In due course the American aircraft was cancelled, so that the Anglo-French project, instead of having a three- or four-year lead, which was the most they could have hoped for and which might well have been overtaken by the greater resources of the American aircraft industry, has ended up with a ten-year lead, and nothing the Americans can do will produce a supersonic civil transport aircraft until well into the Eighties.

Julian Amery likes to think that he rendered Concorde another service. In the final stages of intergovernmental negotiation, based on the sketchy design hammered out by representatives in Paris, a vital matter was raised by the French and the British Treasuries. Both wanted to have a break clause in the contract making either side free to withdraw from the project unilaterally if they thought it was going wrong, was likely to be unsatisfactory, or was going to cost too much. Amery, by his own account, managed to persuade both that it was not possible to tie up the development funds involved and the amount of human brainpower for up to ten years and at the end of the day, or halfway through, find that your partner had dropped out and left the other side carrying the baby. There was no 'break clause' in the Agreement and Julian Amery regards this as the most conclusive decision that he was able to take as a Minister.

The actual signing of the Agreement after the last-minute turmoil involved was little more than a pleasant formality. The ceremony took place in London. It was signed on behalf of the British government by Julian Amery, and for the French government by the Ambassador to Britain, Geoffroy de Courcel. It was only seven short articles long, but it laid down the basic principles. The two countries were to 'develop and produce jointly a civil supersonic transport aircraft' and share equally the management of the project and the proceeds from sales. Two versions were to be constructed, one medium-range and the other long-range, and both versions were to be pursued with 'equal attention'. Four firms were to have the industrial responsibility, BAC and Sud Aviation for the airframe, Bristol-Siddeley and SNECMA for the engines. An integrated management body was to be set up and a standing committee of officials of the two governments would monitor the programme and supervise. To maintain the overall parity agreed, sixty per cent of the airframe construction and forty per cent of the engine effort were allocated to France and the remaining percentages to Britain. Development costs were estimated at between £150 and £170m.

More detailed figures were laid down in the Annexes to the Agreement, which also included a highly optimistic timetable. The first prototype was intended to fly in the latter part of 1966, the first production aircraft was to be ready by the end of 1968 and the aircraft was to enter into service by the end of 1969. But it was to be the development of a new and totally different aircraft. It was to be built within an international framework of control and administration never previously tested and operated, and the best intentions were to prove wildly optimistic. All this lay in the future. The general mood at the time was one of enthusiasm and accomplishment. The European aircraft industry felt that they were in the possession of a potential worldbeater which was going to revolutionize all previous concepts of air travel.

4 Concord(e)

The Anglo-French Agreement had set up an extremely cumbersome framework of managerial and governmental control. It was, after all, the first time that an international consortium had been organized to control a project of such magnitude, and although it looked well enough on paper it had built into it rich sources of discord, delay and duplicated expenditure. I do not want to sound unduly critical. I was part of the machinery myself, and it was as much a pioneering experiment in international administration as was the aircraft itself. Nevertheless we need to consider briefly the pattern of organization, since it survived for many years in spite of minor adaptations and has served as a model for what not to do as well as what to do.

The two management teams, one for the airframe and the other for the engine, were to be completely integrated. Bristol-Siddeley and SNECMA had an eight-man committee which worked remarkably smoothly throughout the operation. On the airframe side there was a chairman, with a deputy from the other country, these appointments alternating; a French technical director with a British deputy; a British finance director with a French deputy; two commercial co-directors and a production director from each firm, adding up to ten.

Each company nominated its top people – the issues at stake were important. General André Puget was the first chairman representing the sixty per cent interest of Sud Aviation in the construction of the plane itself. His deputy was none other than Sir George Edwards, 'boss' of BAC. On our side we were joined by James Harper, who was then Managing Director of the Bristol Aircraft subsidiary, Archibald Russell, deputy to Pierre Satre as technical director, Allen Greenwood, who was sales and service director, and myself as an additional BAC director.

Imposing as this sounds, there was a hidden snag. Both on the French and the British sides, we were all extremely busy men, with major industries to run on either side of the Channel, involved in a multiplicity of civil and military aeronautical projects. We were both engaged at the same time in active competition with the American manufacturers. Both sides also had direct and divorced channels back to their own governments for finance, technical back-up and other ancillary services. Development was to be carried out by the four manufacturers on a 'cost plus' basis.

A standing committee of government officials had been set up to monitor the operation, although one of the Articles did use the word 'supervise'. This was the Concorde Directing Committee, eight in number, with a chairman from one country and a deputy from the other who also alternated every year. The fact that the manufacturers had nominated their very top people to the industry committees obliged the governments to do the same. On the British side we kept it at the level of outstandingly competent senior civil servants, but the French nominated the Directeur Général de Projets du Secretariàt L'Aviation Civile from their Ministry of Transport and the senior officials in their technical and aeronautical services in the Defence Ministry to give it the necessary weight. All these were busy men with immense responsibilities, of which supersonic air transport was only one. It was not long before both governments found it necessary to delegate responsibility to a Concorde Management Board.

At first our own meetings were quite frequent, very formal and extraordinarily tedious. We used to operate with the full United Nations paraphernalia of headsets and interpreters. The technical and production people were allowed to expound at length: they produced brochures for us to read before meetings and then assumed that we had not read them and went on and on with verbal explanations. We were practically designing the airplane round a table, or going through the motions of doing so, which was an impossibility. It was very heavy going and, in the early stages, we used to despair about putting the recommendations together and making an airplane out of them.

We were learning as we went along. There had never been a collaborative major aircraft project in Europe of comparable consequence, and we were feeling our way. In those days neither side was too keen to see the other taking the lead. The French nominally had the senior technical director, with Archibald Russell as the deputy, and we nominally had the senior production man, with Louis Giusta from Sud Aviation as his deputy. I think it would be fair to say that no one considered himself as a deputy to anyone. If his nominal superior wanted to do something that the other didn't like, then the deputy would go through his own management system to his boss and say: 'Look what these chaps are trying to do.' It may be thought that grown men should not have acted in this way, but they were all people of consequence in their own right and had been masters in their own houses for many years. It was asking a great deal of human nature.

We used to meet alternately in England and France, and I would like to pay a tribute here to André Puget for being such a splendid Chairman. As a newcomer and outsider to Sud Aviation himself, he had his own problems, nor had commercial administration or design formed part of his previous experience. He had also inherited intact the powerful and independent-minded team put together by Georges Héreil. Despite all these difficulties he masterminded us all with great diplomacy and was much liked.

Puget had an endearing sense of humour. Some of our early meetings went on interminably, with the committee of directors discussing matters of detail in great depth, which was quite wrong. I used to find it difficult to remain attentive, particularly if someone was rambling on in French, and would switch my earphone control box off. Puget would see me do this and would fire a question at me five minutes later so that I had to hum and hah like mad to get away with it.

Sir George Edwards also played his part as deputy to perfection. He was not a Montgomery to Puget's Eisenhower, for he took enormous care to see that he did nothing to erode Puget's authority both when Puget was in the chair and at those meetings when he was not. Although in terms of sheer experience and engineering skill he was the dominating personality on the committee, he was never overbearing. In the end his views on most subjects were the ones that were accepted. But it was done tactfully, because he had such great respect for Puget as a man.

Sir George had a very quiet way of operating. He would never indulge in a knock-down, drag-out altercation at a committee meeting. If a difficulty cropped up, and I remember there was an early serious difference of opinion about the shape of the wing, he would go down personally to Toulouse to work out the problem with the French, often in the company of Archibald Russell, and persuade them that the BAC solution was the right one. When the matter came up at the next board meeting, there was usually little further argument. I don't think I ever heard him raise his voice. His greatest quality was his ability to convince people of the utter sound common sense of what he was saying. He also had an original turn of phrase of his own, which was sometimes difficult to put across in French. One of the most remarkable men that the British aircraft industry has produced, the part that he played in our long travail cannot be exaggerated.

Some personal incompatibilities were to reveal themselves, although they were the exception rather than the rule. It was surprising that there were not more, in view of the totally different attitudes that the two countries brought to the project. As far as Britain was concerned, the supersonic airliner, although a star turn and the major hope for the industry in the long run, was nevertheless only one element in our economic survival. It would keep our advanced technology at high pitch during a period when our immensely able and experienced aircraft industry was going through a fallow period and would buttress us against losing any more ground to the Americans.

The French attitude was different. De Gaulle had come to power, and they were at last concentrating on the future again, after a bemusing period of military defeat and economic difficulties. The drain on resources caused by the war in Indochina had been stopped, and a solution of the Algerian problem was in sight. France was about to become a great nation again, and her highly gifted technologists and administrators in the aircraft field, with the successful experience of the Caravelle under their belts, saw in the joint project of the supersonic transport plane a means of achieving one element in the Gaullist conception of 'grandeur'.

They had many factors operating in their favour. Caravelle, and particularly the military designs of Dassault, had been conceived and carried through with enormous flair and speed. Organization on the French side was streamlined. Sud Aviation was a nationalized company and, in many respects, an executive arm of government. It had a relatively young, enthusiastic and tightly integrated top team of administrators and designers inherited from the forceful Georges Héreil. BAC, on the other hand, was still something of a conglomerate, made up of component parts which tended to get on with the projects on which they had been engaged before the merger as if nothing new had happened. Sir George Edwards had to spend far more time rationalizing their individualism than he could on the supersonic transport alone. The French were always trying to get down to the actual cutting of metal, whilst the British kept insisting on another engine or wind tunnel test before they would take the next step.

From my own observation there was a greater readiness on the French side to accept a policy decision than there was in Britain. This was possibly a reflection of the lack of cohesion in our own organization. The fact that Satre was the technical director did not mean that Archibald Russell as his deputy would take a decision from him automatically. Henri Ziegler recognized this, and would good-humouredly point out that although the French have the reputation for being undisciplined, our aeronautical engineers deserved it more. Archibald Russell was never able to free himself from the conviction that the French were planning to bring their aircraft industry up to world standards at the expense of Concorde. He felt that they were happy enough to allow the British to decide what sort of paint to use in the final trim, providing the French were allowed to develop the electronic navigation system.

Underneath all the differences in national approach, and indeed contributing to them, was a complete disparity in technical and performance assumptions. In the early 1960s the French believed that supersonic flight over populated areas on short- to medium-haul routes would be both acceptable and possible. Their attitude was totally indefensible and impossible to understand unless that is appreciated. Our whole philosophy, based on the consistently correct thinking of Archibald Russell, was that this was an over-the-ocean airplane suitable for flying at best over very thinly populated countries. There was no way in which we would be allowed to boom our way across Europe and the United States. The full consequences of the sonic boom argument were not yet fully appreciated, but we had it in mind. The French had no sympathy whatsoever with the long-range version, and we had no sympathy whatsoever with their medium-range version. The battle between us in those early years arose from the problem of reconciling these two attitudes.

For a considerable time, the two projects ran in parallel. The greater part of the cost of developing an aircraft is in the flying controls and systems. Archibald Russell was proposing for a period that the British should build a four-engined aircraft and the French a two-engined aircraft, each for their own purpose, but there would be immense savings if each could use the same hydraulic pumps, the same electric generators, the same principles of navigation, the same flightdeck instruments – possibly even the same cockpit layout. There were also radically different estimates between the two sides about the desirable weight of the plane. Even in the preliminary version, the British could not see it fulfilling its airline function below a structure weight of 132,000 lb. The French thought they could produce one to the correct specification with a weight of about 110,000 lb. This meant a constant flow of proposals and suggestions which were analysed and sifted. Initial scheme drawings were made, then more detailed ones, with the size, weight and performance of each component analysed and compared.

Mick Wilde, the Concorde Project Director at Filton, remembers the confusion and cross purposes of those days only too clearly. He says that the French made a fetish of the low-weight version as a form of industrial discipline, although it was realized at a very early stage on the British side that the lighter aircraft was not going to be able to fulfil its function. The process was also extremely time-consuming. It might be possible to get 20 lb off a small component by working a week on it. To get another 2 lb off would take two more weeks, and to remove the last pound could take up to a couple of years.

Wilde remembers that it affected almost everything he put his hand to. A given weight affected the size of the landing wheels and the bearing loads on the runway. Having decided on a certain width and diameter for the main wheel and tyre to match the requirements of pavement strength of the runway, an adjustment in the overall aircraft weight would place a heavier load on them and risked overstressing the runways. It was not just a question of ordering a different sized wheel, because he was also constrained by the geometry of the operating mechanism and the size of the well in the fuselage into which the wheels retract. The weight also affected the amount of fuel that could be carried.

Mick Wilde also recalls the difference in working methods between the two companies. BAC-Bristol had a very closely integrated design organization. For Wilde and his team a design was only as good as the elements that went into it – the structural designers, the aerodynamicists and the systems men. All formed a team. They did not expect the technical director to sketch out a design and get each of these elements right.

The basic design would lay down a set of guidelines for the specialists to work on. Each would then go back to the technical director and raise his objections involving a fatter fuselage or a thinner wing or whatever it might be. Wilde remembers Archibald Russell as being a very flexible man to work under. He would always listen to his experts and was ready to make changes whenever they put up a good case.

When the Anglo-French committee was set up, they found a totally different state of affairs on the other side. Archibald Russell would go to his opposite number, Pierre Satre, with a modification that had been agreed by his entire team. It was the final version acceptable to all. But he found that Pierre Satre did not operate at that level. He spent most of his time at the administrative offices of Sud Aviation in Paris, and the works were down at Toulouse, where Lucien Servanty was very much master in his own house. Satre had to go down the line to his chief designer, chief engineer or chief aerodynamicist to get an answer or an appraisal, and only then was he in a position to give Archibald Russell and Mick Wilde their answer. They found this quite amusing at first, but it soon became frustrating and certainly was a contributory cause of much delay. For all their dash the French worked under a very strongly disciplined hierarchy of administration. Over the years of collaboration with the British it loosened up considerably, but the difference of approach between the two countries was the cause of a fair amount of friction.

There was also the factor that Satre and Servanty on the French side had for some time been rivals. Before the French mergers, Satre had been with Georges Héreil at the Sud-Est company which had made the Caravelle and Servanty had been with Sud-Ouest. Satre was a graduate of the famous Ecole Polytechnique and Servanty, though a gifted designer, was not. The two men kept their distance and Archibald Russell, being a practical man, used to spend more time in Toulouse than he did in Paris and found himself dealing with Servanty more often than he did with Satre. And Russ and Servanty simply did not get on.

It is all a long time in the past now. Russ has retired and Servanty is dead. Nevertheless, there was no doubt that each of them represented personally all the differences in approach, rival technology, rival designs, different administrative methods and clashing temperaments.

In the end, and after many years, Russ won the day. Concorde flies at Mach 2, which he said right from the start was its optimum speed. It is a transatlantic plane with a weight nearly double that under consideration in the early stages, and in all the incredible complexity of its development he got it right. He is a superb aircraft designer, one of the many gifted men to emerge from the Bristol stable, and, in spite of his outspokenness and foibles, much loved.

To give you the flavour of the man, he stood up once at a lunch that Sud Aviation was giving us in Paris and told this story. There was a young man in aviation who was very ambitious but was not making the progress that he expected. He learned that a specialist had developed a means of making brain transplants and went along to see him. He asked about the cost of the operation and was told that this depended on the quality of brain that was wanted. A Red Brick University brain could be provided for about £1000. A Cambridge Tripos brain was twice as expensive. One from an American technical university like CalTech, MIT or Stanford would be at least £5000. But the most expensive of all would be from the Ecole Polytechnique. This cost £25,000. The young man asked what justified this very high charge and he was told: 'Well, it's quite simple. These brains are as good as new and have never been used.'

Much good did come of the meeting of minds. One of the totally new problems presented by the supersonic aircraft was that it had to fly at two speeds, subsonic for taking off and before landing, and supersonic at its operational altitude. This requires quite a substantial difference in trim. In other words, because the aerodynamic centre of pressure moves substantially rearwards as the aircraft goes from subsonic to supersonic, the centre of gravity had to be in two different places at the two speeds to ensure proper aerodynamic balance and control. The British solution to this lay in the aerodynamic camber of the wing. The French accepted this as part of the solution, but it was they who had thought up the idea of pumping the fuel to and fro between the forward and aft tanks in order to provide the necessary trim. It came as a complete novelty to the British, and was immediately accepted as one of the most useful solutions when we first joined with them.

The development of the engine meanwhile seemed to be proceeding smoothly, for there had already been a history of collaboration between Bristol and SNECMA in the early Fifties when the French firm had built the Bristol Hercules sleeve-valve air-cooled piston engine under licence. Moreover engine development is a much more finite process than building an airframe. You can do all the wind-tunnel tests you like on scale models and even on full-sized sections of an aircraft, but it is only when it actually flies that you get all the answers to your questions. The performance of an engine can be accurately measured in a ground test cell. It can be physically weighed. The fuel consumption can be measured and the degree of thrust calculated exactly.

Down at our Bristol engine plant, Sir Stanley Hooker and his team had already developed the Olympus for the Vulcan bomber to the point where it gave almost more thrust than the supersonic airliner needed, and the problem caused by the engine smoking quite badly had been overcome when the Bristol aero-engine plant merged with Armstrong-Siddeley in 1959. For Armstrong-Siddeley had brought with them a particular type of 'vaporizing combustion' chamber which did not smoke. This removed one likely source of trouble with environmentalists.

However, Sir Stanley and his people did have to make a lot of modifications to the Olympus to meet the requirements of prolonged flight at Mach 2. The basic engine was the right size with the right airflow and the right compression ratio, but they had to cope with operating temperatures outside their experience. Not only does air enter the engine at a temperature of nearly 130°C, and the turbine entry temperature reach more than 1000°C, but the whole engine is bathed in an atmosphere of 120°C. The compressors had to be redesigned with titanium instead of aluminium, with heat-resisting steels at the high pressure end instead of ordinary steel, and air-cooled cast blades in the turbine.

The airframe designers required the engine intake to be 'square', for the powerplant underneath the Concorde wing had straight sides and bottom. The accessories had to go in all the corners and be accessible for service purposes. The electrical and hydraulics supply systems had to operate at higher pressures and temperatures. Special lubricants had to be developed, as ordinary oils would not have worked at the prevailing temperatures. All this was highly skilled development work requiring a tremendous amount of proving and testing.

There was one nasty hiccup entirely outside our control, when President de Gaulle decided to veto Britain's application for entry into the Common Market in 1963. Political considerations had played such a substantial part in the project that we all wondered what the future held in store. Curiously enough, this diplomatic bombshell caused hardly a ripple in our affairs. In fact it provided us at long last with a name for the aircraft. Julian Amery recalls that at one of the last meetings he had with his French opposite number, Marc Jacquet, before the de Gaulle veto, the two ministers were discussing what to call the supersonic project. They offered a bottle of champagne to any of their advisers who could come up with the right name. None of them did, and Julian Amery claims that it was he who proposed 'Concorde'. He could not think of any other word which meant more or less the same in both English and French and, except for the final 'e', was spelt the same way. At the time nobody liked the idea very much, but it went down in the minutes as a proposal. When de Gaulle held his press conference to veto our entry into the Common Market, he went on to say that 'nevertheless this should not stop us working together with the British on advanced technological projects such as the Concorde', and when he put his authority behind the name it stuck.

I do not like to spoil an authoritative story, but there is another that perhaps even Julian Amery does not know. At about the same time, Sud Aviation and BAC invited their employees to suggest a name for the aircraft. I have always understood that it was the eighteen-year-old son of one of our people at Filton who first came up with the name Concord(e) because it was one of the few words suitable for the purpose which was spelt nearly the same in French and English. Time blurs most things, and there was certainly a period when the two firms continued to use the word, each in their own fashion, in their brochures and sales promotion material. I think it is true to say that Julian Amery was not too pleased with our failure to agree on a common spelling, but I would be the last person to deny him credit for lodging it officially.

So we had Concorde. What we did not have were any buyers. It must be the only airplane project ever launched without some preliminary understanding with the airlines of what their requirements were and what the market for it might be. It must not be forgotten that at the time it was launched the Americans were in the initial stages of mass production of their subsonic jets, and the world's airlines looked like becoming saturated with Boeing 707s and McDonnell Douglas DC8s with operational lives of at least fifteen years. Britain and France were busily producing medium-range subsonic airliners in the main, with the exception of the long-range VC10, and even Air France and BOAC were distinctly cool towards Concorde in the initial stages. A massive sales operation was put in train by each country, and in June 1963 the Concorde programme had its first breakthrough. Pan American took options on six of the long-range versions, quickly followed by a number of other American airlines.

It was a turning point in the conflict between the rival versions. It took until May 1964 for the two companies and governments to shift emphasis to the larger version, but we were then able to offer the airlines, who had been hammering for a bigger payload than that originally predicted, a trans-atlantic version with a maximum payload of 25,000 lb to carry 118 passengers with adequate fuel reserves. Planned as a one-class airplane, we hoped in 1964 that it would be possible to operate at about ten per cent above the ordinary first-class fare. Passengers were to be seated four abreast, with slightly less space than was becoming usual in the new subsonic types, but our assumption was that if the journey was going to be completed in half the time, slightly less space would be acceptable.

It was also one of our assumptions that the airlines which bought Concorde would integrate this supersonic airplane into their subsonic fleets by running Concorde as a first-class service and operating all-tourist-class airplanes in many of their subsonic services. In our view this would have extracted the maximum economic advantage from each type of airplane. The other calculation we made about Concorde was that as it was designed basically to carry businessmen and government officials, its operation would not be subject to the seasonal fluctuations which affect the tourist trade and that demand for the airplane would remain steady throughout the year. Our estimate was that there would be a total market for about 240 Concordes. Admittedly that was in the days when demand was increasing in the airline business more rapidly than today.

Just when the potential sales situation looked buoyant, a new thundercloud appeared on the horizon. When the development of the larger Concorde was announced, the two governments revised their development cost estimate, which in July 1964 had increased to £200m compared with the original estimate of some £170m. The economic situation was undergoing one of its downturns, and in Britain the major new military aircraft development in which BAC was engaged, the TSR2, had tripled in cost over the original 1959 estimate and was now running at some £250m. There had been serious parliamentary rumblings in January 1964 when the Select Committee on Estimates chose to go through the whole Concorde project with a fine-tooth comb in a highly critical mood. They did not like the way the agreement was working out and they felt that our cost estimating and budgeting procedures were not nearly tight enough. Backbench members on both sides of the House were demanding a reappraisal of the whole project. The French were also starting to have substantial difficulties at government level between the two Ministries of Defence and Transport which had been associated with Concorde from the start, but which had not kept in touch with each other.

Julian Amery was reasonably sanguine that he could ride the gathering storm. As he had said:

> Every aircraft project I have ever had anything to do with has always been changed halfway through, and I certainly wasn't surprised when it happened with Concorde. The development of an aircraft from the drawing board to service was taking about ten years, and in that time requirements change, technical problems arise which mean different solutions, people invent new things and suddenly see the opportunity of making the aircraft better still. It is not something which you make like a box. It is a thing that grows through a steady process of changes. Even when the flying tests begin, new problems arise which have not been taken into account, involving new possibilities of change.

The Conservative government reached the end of its term. A General Election was held in the autumn, and the Labour Party squeezed into power. The result involved Concorde in sudden crisis.

5 The Brown Paper

Labour were elected to power in October 1964 largely because they claimed that they would solve the deteriorating balance of payments situation. Once installed they announced a crash programme of one hundred days of drastic economic measures, negotiated for an American loan of about £900m, and decided to examine all 'prestige projects' both civil and military which they had inherited from the Conservative administration.

We knew that the Labour Party had no particular love for Concorde and would much sooner have spent the money involved on housing or some other social project. One of the first major documents of the Labour administration was the 'Brown Paper' named after its author, George Brown, the new Minister of Economic Affairs. This paper specifically mentioned Concorde. We had expected the TSR2 military plane to qualify for the chop, but this reference to Concorde was an unpleasant and unexpected blow. It has long been stated that the Concorde clause was inserted into the 'Brown Paper' at a very late stage. Julian Amery, who had just lost office as Minister for Aviation, says that this was done under American pressure. It is an interesting story and I had better let him tell it himself:

> Even after the Concorde Agreement had been signed, we were subjected to a ceaseless barrage of advice from Washington to stop the project. Senior American ministers and officials visiting London to see the Chancellor of the Exchequer or Foreign Secretary or the Minister of Defence or

the Prime Minister would, at the end of their conversations on other and larger issues, raise the question of Concorde. Without producing particular arguments, they would make the concerted point that there were very strong feelings in Washington that Britain ought to cancel it. The next day, I would get a note from the minister concerned reporting the conversation and asking what comment we should make. This used to happen about once a month, and I know that these opinions were also being fed to the Labour opposition.

Just before the Conservative government fell from office, I had accepted an invitation to shoot near Paris and, after the change of government, I saw no reason not to take up the invitation. The British Ambassador there, Sir Pierson Dixon, who belonged to the same shoot, very kindly asked me to stay with him at the Embassy. I was there for the weekend and we were out most of the Saturday, intending to stay the night away from Paris. On either Saturday evening or Sunday morning a dispatch rider arrived on a motor cycle summoning the Ambassador back to the capital. He suggested that I should stay in the country as I could easily get a lift back from one of the other guests and, if I got back late on the Sunday night, he would leave drinks for me and we would meet on the Monday morning.

Sir Pierson came into my bedroom on the Monday morning at 8 o'clock and said: 'The most awfully awkward thing has happened. I was summoned back to get instructions to go to see Couve de Murville, the French Foreign Minister, to say that we want to cancel Concorde. I could not get him last night and George Brown is going to make an announcement at 11 o'clock this morning in London. I have an appointment with Couve at nine. I know you were not planning to leave until this afternoon, but as the Press of the world will want comment on this, it might be rather awkward if you were found in my Embassy. Would you mind if my staff got you a ticket and got you back to London as quickly as possible?'

I asked him if it was not very odd that he had been told at the last minute and he also could not understand it. He would have expected to have been asked to go and soften up the French the previous week. When I got back to London I made some inquiries from officials I knew, and the following story emerged.

In the middle of the previous week, Sir Eric Roll, who was George Brown's Permanent Under Secretary, had gone over to Washington on the business of discussing the £900m loan, taking with him a draft of the Brown Paper to discuss with the American government. Someone who claims to have seen the draft that Sir Eric took with him, or a copy of it, had assured me that there was no reference to Concorde in it. When the paper was published on the Monday morning, it contained the sentence about Concorde which looked as if it had been spatchcocked in.

Now the only explanation that I can think of why Sir Pierson had not been informed of developments until the very last minute was that it had not been the intention of the Labour government to put the Concorde clause into the Brown Paper but that the Americans had insisted on it. This would be in keeping with the campaign which they sustained and continued to sustain in favour of cancellation. They did not wish to lend a lot of money to the Labour government to produce civil airplanes that were going to be directly competitive with and embarrassing to the American airlines. When the British government found that there was no break clause in the Concorde Agreement and that the French could have taken them to the International Court at The Hague, where they would have had to pay very high damages to the French Government and probably had the embarrassment of seeing the French go ahead with the project by themselves, they didn't force the cancellation through. I fear that the American President exacted the cancellation of BAC's TSR2 and the HS1154 project as a counterpart.

Julian Amery's account has taken us ahead of events. What certainly happened was that Roy Jenkins, appointed as the new Minister of Aviation, did go to Paris and we at BAC were given formal notice of the fact. Although it was assumed that he was going over to try to cancel Concorde, this was never officially stated, and the government line was that he was going to discuss with his French colleagues a re-examination of the viability of the whole project.

It happened, quite by chance, that I was in Paris on an ordinary business visit the day before Jenkins arrived, so I took the trouble to go round and see a very old friend, Paul Simonet, who was the Secretary General of Sud Aviation, to discuss the matter. He told me very clearly that he could not see the mission getting very far with the French government. The whole business about taking the matter to The Hague court came up later and certainly happened above the heads of the two firms. There is no doubt that the French took an exceptionally robust line over the whole matter and certainly made it clear that the contract did not provide for a unilateral break by one partner at any time. So Julian Amery had worked better than he knew two years earlier, although, as he himself admits, what he had in mind was to prevent a French cancellation.

The government side of the story has yet to be told in full. The Attorney General was asked to examine the legal situation to determine whether, under international law, the French would have a case, and there is little doubt that he came to the conclusion that they would. The crisis lasted for three months, until the middle of January 1965. During that time, government officials reduced their contacts with the French to a minimum and we in BAC maintained a low profile and kept our exchanges as quiet as possible. We had just reached the stage of constructing the prototypes, and all work was slowed down for a time, with delays on the purchasing and subcontracting side. The programme by then was so large that there was no way of stopping activity altogether. One good result was that, with the two governments squabbling, the two airframe companies came closer together than ever before in self-defence. My recollection is that there was very much a 'business as usual' atmosphere. There were no great arm-wavings and cries of despair, and we just got on with the job as best we could.

Faced with a difficult economic situation the Labour government wished to cut back on its expenditure, but at the same time it needed to carry the unions with it. The trade unions were violently and noisily opposed to Concorde's cancellation, as this would create redundancy. Clive Jenkins' white-collar union was particularly incensed. The enthusiasm of the unions for Concorde has always been a great asset.

There was an interesting parallel in France. Although the French labour unions wield nothing like the power of those in Britain, they rose to a man in defence of Concorde. A socialist deputy from Toulouse named Eugene Montel went to London to talk to members of the government and fellow unionists, and proclaimed on his return that the British would continue to build Concorde and what the Labour government sought were economies and not cancellation.

By February 1965 we knew we had been reprieved. The Labour government announced that all three of the major military aircraft development projects were to be cancelled. Hawker Siddeley suffered the loss of the HS1154 and the HS681, a vertical take-off fighter and a short take-off transport, and BAC lost the TSR2. This was a bitter blow for the British aircraft industry, but at least we still had Concorde. Also in our favour was that after the election, the government had set up the Plowden Committee to study the aircraft industry, and its report had recommended joint ventures with European partners as the best basis for future activity in the British aircraft industry. We signed one agreement with France in May 1965 for the BAC/Dassault–Breguet Jaguar supersonic tactical fighter, together with a proposal to study an advanced swing-wing fighter-bomber with Dassault. Thus reinforced, we felt a little more sure of our ground and concentrated again on our principal obligation.

6 Cutting Metal

It might be supposed that after the 1964/5 cancellation scare had run its course the Concorde project was away and running. But our difficulties were by no means over and a whole new series of problems were to arise. This is perhaps a good moment also to catch up on two parallel events which affected the general situation: the merger with Rolls-Royce and BOAC's reduction in the number of VC10 aircraft on order.

In 1964 the original Bristol Aeroplane Company still existed, although its airframe subsidiary had been absorbed into BAC. Then, in that year, the Bristol Aeroplane Company itself was bought out by Rolls-Royce, who at the same time acquired the Hawker-Siddeley shares in Bristol-Siddeley. Thus the whole of the big Bristol-Siddeley engine company was taken over by Rolls-Royce, who became monopoly aero-engine suppliers to the United Kingdom. This made very little difference at Bristol to the supply of Olympus engines for the Concorde, because the Rolls-Royce management at Derby tended to operate the Bristol engine works very much as a separate entity. In the course of time this separation became accentuated as the management at Derby concentrated its activities to an increasing extent on the RB211 engine which was eventually to have such a dramatic effect on Rolls-Royce's fortunes. In buying Bristol, Rolls-Royce also inherited a twenty per cent interest in BAC, which made them somewhat odd bedfellows. In due course, this stake was bought out by Vickers and GEC, who then each owned fifty per cent of BAC.

During the same period of the Sixties, a serious blow was

suffered by BAC with the decision taken by BOAC to reduce its order for Super VC10s. This was the airplane that in due course was to make a high proportion of BOAC's profits and was to become probably the most popular passenger airplane that travellers have ever flown in. However, at that time a different view of its potentialities was taken by BOAC. Sir Giles Guthrie had been appointed Chairman, with a firm directive to put Britain's overseas airline on a profitable basis. Sir Giles, a pleasant character with whom I have always got on extremely well, received clear-cut assurances from his Minister that there would be no government interference with his day-to-day running of the airline and, after an exhaustive study, he had come to the conclusion that the aircraft was less profitable to operate than the Boeing. But by that time BOAC had thirty Super VC10s on order.

The VC10 was one of the main civil aircraft projects that Vickers had brought into the merger when BAC was formed, so that when BOAC announced that they wished to reduce their order of thirty Super VC10s to seventeen, it came as a severe shock. As things turned out, the VC10 proved to be not only a highly popular passenger airplane but also a highly profitable one, so that the original reason for the cancellation and the alternative investment in Boeing 707s were invalidated. It was I think another of those cases which tend to sadden people in the British aircraft industry, when decisions are taken which appear at the very least to be against the national interest and later on turn out to have no commercial basis either. I would apply the same argument to the cancellation of the TSR2 and the switch to the American F111 and then the Phantom.

While all this was going on, BAC was engaged in parallel discussions with BOAC about Concorde. This involved our Bristol team, as distinct from the Weybridge team who were dealing with BOAC on the VC10. Considerable pressure was being exerted on BOAC to place an order or to take out options on Concorde. However they remained adamant that they were not going to do either until they had a much clearer idea of the capabilities of the airplane and of its economics and performance. Here was yet another example of BOAC, probably quite properly, maintaining its commercial independence. But who coordinates and directs the national interest when one arm of government decides on a massive capital investment in a new aircraft development while another arm of government displays no great enthusiasm for it?

Another factor which gave us some cause for concern was internal rather than public. The Olympus engine for Concorde had not only been designed for the Vulcan bomber but, in its improved form, was to power the TSR2. And TSR2 had been wiped off the map. This meant that the immense development costs for the improved engine, a high proportion of which would have been carried on the defence budget, was now to be marked up exclusively to Concorde.

Although on the surface the loss of the TSR2 contract and the decrease in the number of VC10 orders would seem to leave us with more space, more time, more men and more energy for Concorde, and it might be thought that it would give us a chance to go ahead faster, this was not so. For the demise of the TSR2 in 1965 (only two and a half years after we had started on Concorde) meant the loss of the big fund of engine experience that it would have built up for us by now. This in turn meant that a whole new dimension of cost, time and effort was added to the Concorde development programme at a stroke.

As I have explained in previous chapters, the Olympus engine seemed easily adaptable to Concorde. We did not think the cost of 'civilianizing' it would be high, and as I have already pointed out the Olympus gave almost more thrust than a supersonic airliner seemed at the time to need. But we were premature in thinking that engine problems would be less than airframe ones. New complications in the engine field began to mount.

Very considerable extra work was required to keep the engine output in line with the ever-growing weight and performance requirements of the Concorde airframe, and the ever-increasing range of airworthiness and safety requirements to be fulfilled in translating the initial technical concept into ultimate airline operational reality. All this extended far beyond the engine itself to embrace the complete powerplant, and also necessitated a correspondingly larger number of development engines. What had been initially regarded as a comparatively simple and inexpensive task ultimately accounted for nearly half the total Concorde programme costs.

By this time, both firms were starting to 'cut metal' for the two identical Concorde prototypes, 001 in France and 002 in Britain. There comes a time, if any progress is to be made at all, when designs have to be frozen. Jigs and tools have to be made to shape and produce the components and the whole complicated process of transferring designs on sheets of paper to interlocking hardware has to begin. In spite of the almost total lack of response to the French overland medium-haul version, they still clung to their original conception as a necessary element in development. Neither of these two prototypes were at this stage long-range aircraft. They were never going to be non-stop North Atlantic airplanes and were essentially basic test vehicles from which to develop the ultimate production version.

In 1965 the standing committee of the two governments, of which J. A. Hamilton was the principal representative on the British side, and Directeur Général Gérardin on the French, decided that for the first and only time in this long-drawn-out process, they had to knock our heads together. With the British long-range concept finding initial favour with the international airlines, the committee decreed that the next stage – the pre-production planes 01 (UK) and 02 (France) – were to be 8½ feet longer and would have a take-off weight of 367,000 lb. They would provide space for up to 136 passengers.

This gave a lot of satisfaction to Archibald Russell. He tells me that for years he had been carrying around a piece of paper on which was written his basic conception of the Bristol 198, the forerunner of the Concorde agreement on the British side, which was formulated in 1961 and was for a 136-passenger, long-range plane. He had never really wavered from a 385,000 lb take-off weight – an empty weight of 135,000 lb – but there was still another stage to go before he finally achieved it. In the end Concorde was designed three times. Complete sets of drawings were required for the two prototypes, for the two pre-production models and for the eventual production versions. Even in retirement, Russ never ceases to claim that, if his original version had been adopted, Concorde would have been in the air in half the time.

It must not be forgotten that we were pushing out the known limits of technology. This is something that might have been done more quickly with one firm working on its own. Political considerations, the problems of cost, continuing divergences of opinion about the final shape of the plane, and the more intangible element of national and professional pride all played their part in the complicated situation in which we were involved. Even though the French, with some reluctance, had now accepted the concentration on the heavier long-range model, they were still convinced that they could produce it with a substantially lower take-off weight than the British design side considered possible. The French, for quite a while, were convinced that they could produce Concorde to the new requirements with an empty weight of 108,000 lb, whereas Archibald Russell's figure was 135,000 lb.

The trouble here is that, in terms of performance and in comparison with the American conventional subsonic airliners, Concorde is a poor subsonic plane. It has a lift/drag ratio, to be technical for a moment, of about 9 instead of the 16 which is the figure for the Boeing 707 and McDonnell Douglas DC8. This means that Concorde has to have something like fifty per cent more reserve fuel in order to cope with the problem of congestion and stacking over the destination airport if there are delays in bringing aircraft in to land. Providing that the Boeing has 20,000 lb of fuel to spare, it has nothing to worry about. Concorde needs 35,000 lb of reserve fuel to cope with this situation, and the difference has to be deducted from the payload of the airplane. This is the sort of requirement that the potential customer airlines were always nagging us about, with the result that the overall weight kept increasing, the payload kept going down and more and more demands were made on the engine manufacturers to produce more thrust. Over the whole history of Concorde, something of the order of 50,000 design drawings alone were required to cope with the constant demands for modification and improvement.

All this was being done under a form of organization which proved inadequate to the purpose. The number of lines of communication was bewildering. BAC communicated with Sud Aviation; SNECMA collaborated with Bristol-Siddeley; but each of the four firms communicated separately with its own government, by whom it was paid, and the two governments communicated with each other through their Concorde Management Board. This was the pattern that had been laid down. This was the form of administration that functioned, more or less, during the formative years of the Concorde project. It had been a bad solution to a completely novel requirement, and we have learned the necessary lessons from it.

As it was, every proposal made by a Frenchman was criticized to the death by the British and every British proposal was torn to pieces by the French before some sort of compromise was reached. There has never been an airplane made where there has been such criticism of basic assumptions. This was not all necessarily bad from the point of view of the end product. The fact that everything was scrutinized in such great detail by both sides did mean that by the time we reached agreement on a piece of design or a modification it had been examined much more thoroughly than has probably been the case in any other airplane developed to date. I think this shows in the very high-class performance and reliability that we have in the airplane today.

On the engine side, the demands of the designers and the airline operators kept them continually on the hop. The number of test and development engines required for the programme was several times what had been anticipated. They had to keep redesigning the engine to meet the demands for additional thrust. Stanley Hooker and his people had to provide for a lot of access holes in the carcass of the engine through which to push boroscopes in order to see what was happening inside without taking the whole engine apart or out of the aircraft. These had not existed on the engines, and you cannot make a hole in the side of a turbine engine just like that. There was a very large amount of additional testing under simulated supersonic conditions where the air on the test bed had to be heated to 120°C so that the engine could run for long periods under conditions that simulated high-altitude flight. No one foresaw the extent to which this proving programme would grow over the years. In the end, the engine had to be pushed to the absolute limit of capacity for its size, although, like everything else on Concorde, it works beautifully.

A further difficulty was that BAC was not able to make a decision which involved the spending of money without comprehensive justification. I recall that any modification that was going to cost more than £10,000 had to be approved by the office of the Concorde Director General in the Ministry. This in its turn often led to problems of communication and delay. £10,000 may sound a lot of money, but this is not so in the aircraft business, with all its problems of development, improvement, tooling and production. This was the element in Concorde which made it much more like a defence contract and, although we had plenty of experience of this sort of thing, it does sap energy in the end and removes from a company the kind of freedom of action that its senior management is paid to take. The French were much more open-handed in this respect. It is also true to say that they had many fewer officials involved in the Concorde operation than we did. Sud Aviation was a nationalized industry and control was much more direct. Gérardin, the French Director General, controlled a streamlined team which accepted the technical and financial responsibility for several projects including Concorde. They tended to accept the advice they received from Sud Aviation, whereas with BAC it was always a matter for negotiation with our officials.

This is not intended as criticism of our own senior man at the Ministry, Jim Hamilton, whom we were very lucky to have. He was one of the few trained engineers who was later to attain Deputy Secretary level, and we were always able to talk to him on the basis of equality of knowledge. Nevertheless his team was subjected to the usual frequent shifts and postings of the British Civil Service, and he always had the Treasury on his neck, with their well-known aversion to spending any money at all, particularly on prestige projects.

There was another overriding element in our relationship with Sud Aviation which had really nothing to do with technical differences at all. There was, certainly at top level, a suspicion among the British that the French were trying to secure for themselves the prime position in aerospace in Europe, and no doubt a suspicion in France that the British were trying to do the same. For the first fifteen years or so after the war, the British had a far bigger, more competent aerospace industry, with a capability that went across the board in airframes, engines, undercarriages, and components, which was not matched in any other country in Western Europe. So there was a certain amount of national jealousy between the French and the British on that score. The French also found it more difficult than we do to deal with the United States. We find it pretty easy to get on with the Americans. Their thought processes are in a lot of ways quite similar to ours.

All this spilt over into the question of language. Almost all the Frenchmen at Sud Aviation spoke enough English to be able to conduct their business with us in that language. Over the course of the years, I regret to say that few people in BAC got round to speaking enough French to conduct business even for two or three minutes. Increasingly over the last thirty years or so, English has become the international language in which people of all nationalities converse, particularly on aviation and technical subjects, but for centuries French had always been the language of diplomacy, and they do not like to see it ousted. I remember the telling remark that President Pompidou once made in an interview on television when he said: 'English is the language of the Americans.' This was another prejudice we had to try to overcome.

It had become clear, by the turn of 1965–6, that the cumbersome administrative pattern set up to control the Concorde must make way for a smaller and more efficient system. The joint board of ten directors of the two airframe firms met barely once a quarter. The intergovernmental committee was hardly more assiduous and had taken to delegating increased authority to its technical and administrative subcommittee. This was therefore upgraded and renamed the Concorde Management Board. Jim Hamilton went on it on the British side, with the formal designation of Director General, and on the French side M. Gérardin, who had been having increasing difficulties with his dual loyalties to his Ministries of Defence and Transport, was succeeded by Ingénieur Général Forestier, who reported equally to both ministries.

The two airframe firms also succeeded in coming to an arrangement which did away with many of the personal antipathies and problems of the past. The Committee of Directors of the two firms created a Concorde Executive Committee, headed by Dr Archibald Russell and the number two at Sud Aviation, Louis Giusta. This made an enormous difference. Louis Giusta and Archibald Russell were as different as chalk from cheese but they got on well. The improvement in Anglo-French relationships was immediate. They met frequently, every ten days or so, and all the technical problems, both French and English, came up to them for joint decision, which they took with a minimum of friction. The combination worked extraordinarily well.

However, we were to have one serious casualty. Over the first four years of the joint project, a tremendous part had been played by General Puget as chairman of the joint Committee of Directors in keeping the whole project going, in spite of all the differences of outlook, the personality clashes, the technical confrontations and the political infighting. In January 1966 there was a total reappraisal of the estimated development cost of Concorde. To the general alarm it had doubled again over the previous figure and was now estimated at £500m. Curiously enough, this caused relatively less uproar in the British Parliament than had the crisis at the advent of the Labour government in 1964. The estimates were approved by the Concorde Directing Committee in May and by both governments in June.

This time the more serious rumblings came from France, although they were more muted and less raucous than we tend to be in Britain. Nevertheless the French Ministry of Finance had become highly critical of the cost overruns on Concorde and this undermined Puget's position. He used his immense prestige to obtain the French government's approval, but he had played his last card. Quite unexpectedly, later in the year he was nominated Ambassador to Sweden and was replaced at Sud Aviation by Maurice Papon, who up to then had been Prefect of the Paris Police.

The Papon appointment was curious considering his previous experience. It was difficult to see any connection between the work he had been involved in up till then and the aviation business. Nevertheless, he was offered the Presidency of Sud Aviation, which he decided to take. It is not surprising with his background that he turned out to be a very different head of an aviation company from anyone we had been used to. But by this time the Russell–Giusta team was working well, so we found that we had very little to do with Maurice Papon who only stayed a couple of years.

The escalation in cost estimates certainly resulted in much closer ministerial and Treasury control of our activities and expenditure. The very early design stage of Concorde had not cost a great deal of money. Project and design groups form small, highly skilled teams, and although they are pretty well paid there are not many of them. When you get into development as opposed to design study and start laying target dates for first flight, entry into service and so on, the spending starts going up in a very sharp curve. You lay out money on planning, tool design, detailed drawing, building jigs and tools and purchasing your raw materials. That is when the money really starts to get spent. There is a fundamental difference between monitoring costs that have actually been incurred and estimating costs yet to be incurred. What happened with Concorde, and it is common to all long-term technological projects, was not that we were greatly exceeding our budget year by year but that the work still to be done expanded all the time.

When you are dealing with a major development job, you are, by definition, exploring the frontiers of knowledge. It is improbable, to say the least, that your estimate for what you are proposing to do will hold right through the contract. It is simply not possible to give precise estimates when you are involved in major and complex development jobs. Everyone complains about it – the Treasury, ministers, civil servants and the shareholders – but those are the facts of life.

We made regular returns to the Ministry on standard forms in great detail to show what we had spent under various headings. Each year, at least, we had to re-estimate the total cost to completion, which we did at the current level of costs and prices because it was the government's decision to make its own estimates of inflation, which at this stage had not yet reached the acceleration of the 1970s.

In BAC, most of the initial work was done at the Filton plant. We knew how many draughtsmen, how many technicians, how many fitters and how many assembly workers were going to be on the payroll in a particular year. What we did not know was whether twenty per cent of the work they did would be abortive and need to be done again the following year in a different manner, or whether we would fail to complete as much as we thought we were going to be able to in any given year. With the best will in the world we could not know to what extent targets were going to slip so that although we had spent the money we might not have reached the particular stage that was planned and so have to spend some more.

This was very much like the system of defence contracts, with which we were thoroughly familiar. Although it sounds cumbersome, it worked. The British government were quite fast paymasters once we had gone through the drill, so that we were never short of cash. Sud Aviation used to be surprised and jealous. Very often months went by while they waited to be paid, although they did not have to go through the complicated checks and balances to which we were subjected.

In December 1967, the Concorde project reached its first milestone, when the French-assembled prototype 001 was rolled out of its hangar at Toulouse, followed shortly thereafter at Filton by the British-assembled 002. Neither of them was much more than the shell of a test-flying aircraft, but at Toulouse it was possible for the new British Minister of Technology, Anthony Wedgwood Benn, to contribute one of the more pleasant courtesies of the whole project when he announced finally that the name of the aircraft was to be spelled with an 'e'. 'The letter "e" means Excellence, England, Europe and Entente,' he said.

Before and during this time, another set of problems was being resolved. A great many of the components and control systems were of necessity put out to suppliers and subcontractors. This is perfectly normal in the aircraft business as it involves highly specialized work, particularly in something as sophisticated as Concorde. However, the two firms and their governments had become involved in a series of unseemly squabbles as to how this work should be parcelled out. Obviously it was worth a great deal of money and, for reasons of national prestige, competing manufacturers on each side of the Channel were lobbying BAC, Sud Aviation and the ministries through their MPs and others, so that they could get a share of the action.

Sud Aviation and BAC had to write specifications of their requirements, and these had to be submitted to every conceivable supplier in France and Britain to await their proposals. We then had to take into account the experience of the firm, the technical merit of their proposal and its cost, and the technical back-up needed in operation and servicing facilities. We had to make a first, second and third choice, and these would be submitted to the two ministries for them to advise us which company should get the contract.

The French had an ineradicable objection to any work other than that absolutely necessary going to American subcontractors. This was a political rather than a practical consideration, as the Americans already had some of our needs in stock and we could have bought what we wanted off the shelf. However problems are bound to arise with international collaboration. It is no good glossing over these points, which are all part of the story of Concorde.

7 Testing, Testing

In 1956 the Supersonic Transport Aircraft Committee was set up in the UK to study the prospects for flying civil aircraft at supersonic speeds. On 2 March 1969, exactly ten years less one week after it had published its findings, Concorde 001 took off for the first time from Toulouse with Aérospatiale's chief test pilot, André Turcat, at the controls. This first flight lasted barely half an hour. On the following 9 April, Brian Trubshaw made the maiden flight on the second prototype 002 from Filton. The great dream had come true.

These were exciting days for both firms, but there was much yet to be done before it would be shown that Concorde could operate successfully at over twice the speed of sound or around 1350 miles an hour, at up to eleven miles above the earth and carry over one hundred passengers on the important airline routes of the world. It was seven months after the first flight that Concorde first flew at speeds greater than that of sound. Both prototype aircraft were in the air on this important day but the barrier was broken by Turcat when a minor technical problem developed in Trubshaw's aircraft. From there on we were in unexplored territory as far as civil aircraft were concerned, pushing the flying speed and altitude well beyond the subsonic transport experience. Each step was taken carefully with infinite checking of performance and response of the aircraft, only building up gradually to Mach 2.

When we were able to check at Mach 2 all of the theoretical calculations of performance which we had made many years earlier, we found almost incredibly that these predictions were within one per cent of the measurements. This was a quite extraordinary achievement, unique in the history of aviation, and a tremendous tribute to the skills of the designers and engineers in both Britain and France who met the greatest challenge of their careers. However, as in all fields of high technology, there were problems to be surmounted.

I am no technician. I will try to explain therefore in layman's terms what people like Sir Archibald Russell, Doctor Bill Strang, Mick Wilde and Sir Stanley Hooker, with their French opposite numbers Pierre Satre, Lucien Servanty and Michel Garnier, succeeded in doing. This is their chapter.

Strangely enough the first point they make is that one of the most difficult problems was to design a supersonic airplane which would behave absolutely normally at subsonic speeds. They had determined that the slender delta wing shape was necessary to meet the supersonic requirements and they had to make this wing behave in all respects like the conventional subsonic civil aircraft such as the VC10, the Boeing 707 or the DC8. Apart from the obvious preference of pilots to have these familiar handling characteristics, the code of safety requirements, laid down by the airworthiness authorities throughout the world, had evolved around this kind of aircraft. Concorde was therefore required to meet the same set of rules – to land safely at airfields in crosswinds of thirty-five knots, to use the same automatic approach aids and to work with the air traffic control systems as they exist today.

The capability to achieve conventional subsonic behaviour is not something that can be corrected in the late stages of design; it must be taken fully into account when the slender delta wing is first conceived. The secret lies in something called 'anhedral' – the feature which we see as the droop of Concorde's wing tips towards the ground. This peculiar shape is not what production engineers would like to build. They would have much preferred a nice flat wing which would have been much easier to manufacture – but they recognized the requirement and overcame all of the difficulties in meeting it.

Details of this complex shape had to be worked out using small-scale models of the airplane tested in wind tunnels. Even for this subsonic design requirement, many wind tunnels throughout Great Britain, France and elsewhere were brought into the programme, and the design was checked and cross-checked before being finally adopted. Of course this same wing shape needs to behave equally well in the so-called transonic regime, that is the transition from subsonic to supersonic speed, and then on through to the cruising speed of Mach 2. For these operating conditions other specialized wind tunnels needed to be used – in all some thirty to forty separate installations.

This phase of the early design work required a great deal of attention and cost money and time. The effort was well spent. Right from its very first flight Concorde has been praised by all the regular airline pilots who have operated it, most of whom had never sat in a supersonic airplane before. We succeeded in giving the pilots the impression of flying in a totally conventional aircraft in spite of the unconventional shape which surrounded them. As a result, the conversion of the normal airline pilot to Concorde flying will require no significant increase in the familiarization training. In this respect, our achievement exceeds that of the military airplane designer who relies on very specialized pilot training programmes.

Although the wind tunnel testing of models came out with all the right answers for the subsonic and supersonic cruise behaviour of the wing, they were not fully correct in predicting the transonic characteristics. It is more than usually difficult to get the aerodynamic loading in a transonic wind tunnel to be fully representative, and this was particularly true in the 1960s when Concorde data were being collected. Experimental and theoretical techniques have improved since those days, but Concorde exhibited one of its significant problems when first flown at these speeds. We ran into the effect of flexibility of the wing structure, that is the tendency of the wing to change its shape when certain kinds of air loading are applied to it. The delta wing on Concorde has all of its control surfaces distributed along the rear (trailing) edge of the wing, and these surfaces are used to control the aircraft's attitude and to change the direction of flight. If the pilot wishes to roll the aircraft in turning to the left, for example, he will apply 'up' control on the left wing and 'down' control on the right. At transonic speeds all of the load from these control movements will appear on the control surfaces themselves and, being located at the trailing edge of the wing, these applied loads will twist the wing, particularly towards the tip.

We found that the load arising from the twisting of the wing counteracted the load applied by the control itself and the aircraft did not roll into the turn as intended. It was therefore necessary to modify the way in which the controls moved in response to the pilot action, making the inner controls do most of the work since these are attached to a less flexible part of the wing. This problem, which was successfully overcome, was typical in that it required total collaboration of the aerodynamic, structural, systems and production engineers in both countries.

Having got the wing and its controls right, the next major problem arose from the powerplant as we extended the flight envelope out towards the Mach 2 cruising speed. The powerplant on supersonic airplanes is not just the engines but also the intakes which feed air to the engines and the nozzles which exhaust the air. The Olympus engine, which is the responsibility primarily of Rolls-Royce, is relatively conventional and requires to be fed with air travelling at about one half of the speed of sound. The air entering the intake does so at the cruising speed of the aeroplane, that is at twice the speed of sound, and it is the task of the intake to slow the incoming air, before it reaches the face of the engine, from Mach 2 to Mach 0·5 – a change of 1000 mph in the intake length of 14 feet. The task of slowing down air to this extent is much more difficult than on subsonic aircraft, where the deceleration of the air is only about 200 mph. The supersonic incoming air must first be passed through a series of controlled shock-waves to bring it just subsonic, and the complexity originates from the need to have carefully shaped surfaces inside the intake from which these shock-waves are generated.

Because the intake needs to work over a wide range of flying speeds from take-off to cruise, these inner surfaces of the intake on Concorde need to be variable in their geometry, and a special control system has to be used to ensure the correct geometry in all conditions. Concorde uses rectangular-shaped intakes so that the movement of these inner surfaces is more readily achieved. The intake produces almost seventy per cent of the gross thrust of the power plant in cruising conditions, which shows the importance of obtaining a correct geometry. We met a great deal of trouble in obtaining the proper form of control system for the intake moving surfaces, not for straight and level cruising flight but in the remote cases such as apply if one engine were to fail. Such an incident would disturb the motion of the aircraft and so disturb the flow into the other three intakes.

The safety authorities lay down stringent regulations governing the behaviour under such remote conditions and we spent a great deal of flight test time in obtaining a totally satisfactory intake and intake control system. Our difficulties in this respect were increased because we were unable to find any satisfactory ground facility which would give all the necessary answers, but we have finished up with the most docile combination of intake and engine which has ever been seen on a supersonic aircraft.

The exhaust nozzles on supersonic powerplants are also much more complicated and important than those on subsonic airplanes. In fact, Concorde needs two nozzles for each engine. The inner nozzle carries the air which passes through the engine; around this inner nozzle lies a secondary nozzle in which is mixed not only the engine air but the cooling air which has passed around the outside of the engine. Because of the wide range of speeds and altitudes at which the airplane operates, both of these nozzles have to have variable geometry and be automatically controlled at all times. The air which has passed through the engine has a pressure which is fourteen times that of the ambient condition in high-altitude cruising flight, and again it is most important that this high-pressure air passing through the primary nozzle be mixed efficiently with the bypassed air and expanded out through the secondary nozzle with the maximum efficiency – nearly thirty per cent of the gross thrust is generated in this region. Added to all of this complication is the need to combine within the two nozzles the thrust reversing function for which all of the engine air has to be diverted forward to shorten the landing run.

We have had many problems in this nozzle area on Concorde, leading to three completely different designs from the prototype first flight to the production version. A not insignificant reason for these expensive changes is the difficulty of perfecting nozzle design in ground test facilities. To be sufficiently accurate, ground testing requires that the hot engine air flow and the cooler bypassed flow be properly represented and, at least as important, that the external air flow over the nozzle correctly simulates that occurring on the aircraft. This external flow, which on the aircraft has passed over the wing, is almost impossible to simulate on the ground, and our initial nozzle designs were wrong for this reason. The long and expensive development path on nozzles has finished up with a very lightweight and efficient unit. The thrust reversers are not only used on the ground but are also used in subsonic flight to replace the air brakes which are often seen extended on subsonic jets to slow the aircraft down.

Considering the extensive nature of the international collaboration which was called for in Concorde design and development, the number of problems such as those mentioned above has been remarkably small. The collaboration has required that about a thousand English-speaking engineers should understand and work well with a thousand French engineers, and that they in turn should satisfy the safety requirements laid down internationally and interpreted by the British and French airworthiness authorities. In itself this task was difficult enough, but it must be remembered that the airlines have a great deal of influence, particularly in the layout of the flight deck, and passenger cabin. Through the 1960s we had to co-operate with sixteen airlines who held options to purchase Concorde, each of whom had their own ideas on the layout of these areas.

The flight deck layout was our most difficult collaborative exercise, since it is well known that hardly any two pilots can agree on the precise layout of instruments and equipment. On Concorde, where flight deck space is at a premium because of the need to obtain a very slender shape, the agreement with sixteen pilots was much more difficult. It is a tribute to the patience and quality of all concerned in this work that Concorde's flight deck today is considered by all pilots who have flown in it as being entirely satisfactory, with most excellent visibility for the critical take-off and landing manoeuvres.

The famous droop nose of Concorde is largely responsible for the achievement of good visibility. The British had earlier incorporated this feature in an experimental aircraft known as the Fairey Delta 2, and persuaded French colleagues of its virtues when Concorde was first schemed. By having a moving nose the conflicting requirements of good visibility near the airfield and a smooth streamlined shape for supersonic flight can both be satisfied. On the Concorde prototypes, when the nose was raised after take-off there was no forward visibility available. Subsequently, at the insistence of the pilots, transparent panels were provided and the crew have all-round visibility in all conditions on the production version. This illustrates one of the many valuable lessons that were learned by flying the two prototypes well ahead of the final production aircraft. For Concorde, with its many novel features, there could be no substitute for having practical flight experience. However, taking a balanced view, it is obvious that most of the original design decisions stood the test of experience, and the fact that today's production aircraft are so like their prototype predecessors bears witness to this.

One crucial and correct decision was to design the aircraft

to fly at about Mach 2 and therefore to be able to use more or less conventional aluminium alloys as the prime structural material. The outside surface of Concorde in cruise is hotter than boiling water, due to the friction generated by the passage of air. We knew that aluminium alloys could sustain temperatures somewhat above one hundred degrees centigrade, but knew little of the long-term properties of titanium and nickel alloys which would have been necessary for flight at Mach 2·5 or Mach 3. Our American friends spent a great deal of money in attempting to design a civil supersonic transport for this higher-speed regime, and eventually had to admit that it was too expensive even by their standards.

Knowing that aluminium alloys could work on Concorde was, however, a far cry from proving it. It is conventional practice in aircraft design to manufacture large structural components and to establish by applying loads in a laboratory the exact strength which has been achieved, both under sustained large loads and under the repeated application of smaller loads such as occur each time that pressure is applied inside the passenger cabin. Such tests become much more difficult to perform when it is necessary also to raise the temperature of the structure under test in the laboratory, simulating the effect of air friction at supersonic speeds. Testing hot structures is difficult, but in the case of Concorde the structure is not uniformly hot – for example the inner wing structure which is immersed in the fuel is much cooler than the outside surface in cruise. This difference of temperature itself creates additional loads due to thermal expansion, and these loads are repeated during each flight of the airplane. Repeated loads due to thermal effects are additional to those arising from cabin pressurization or air loading, and can cause cracks to occur in a structure in the long term. Such failures are called fatigue damage.

To test the resistance to fatigue damage on Concorde we had to build a complete aircraft and test it under the cyclic load conditions generated by thermal effects, pressurization and air loads. An enormous ground facility for this work, with massive heating and cooling facilities, was built at the Royal Aircraft Establishment at Farnborough, and the complete Concorde specimen has been under test there for many years. These tests will go on, always making sure that the test specimen is subjected to a much greater number of cyclic loads than any aircraft is likely to meet in commercial service.

Another complete Concorde was subjected to the very large loads necessary to establish the ultimate strength of the prime structure. Again these tests needed to have a proper representation of the temperature effects, and a special facility for this purpose was provided in France. In both the static strength tests and the fatigue tests we might have tried to use smaller components of Concorde rather than the complete aircraft, but by so doing we would have run the risk that the results would not be interpreted correctly on to the complete aircraft. We judged that the full airframe tests would leave us in no doubt, but such precautions underline the claim that Concorde is the most thoroughly tested civil aircraft that has ever been flown.

Concorde is built by joining up over twenty separate structural components. The join-up process takes place on two assembly lines, one in Toulouse and one in Bristol. However, the structural components for each assembly line are produced at either one of the Aérospatiale or one of the BAC factories, and there are eight of these factories producing structural components distributed throughout the two countries. It may seem remarkable that these components built in so many different places should fit together without problems when they meet on the two assembly lines; perhaps even more remarkable when it is appreciated that the French worked in millimetres and the British in inches. In fact, the many hundreds of join-up processes involved have never created a significant problem.

This achievement comes from applying tremendous attention to producing every drawing in both dimensional units, in cautious, painstaking planning and quality control by the manufacturing departments, and by achieving agreement on the many thousands of standard items such as rivets and bolts which must be used on both sides of the Channel. None of this could have worked without the total cooperation of the British and French engineers involved, many of whom have dedicated over a decade of their lives to this task.

8 New Management

At the end of the Sixties certain major changes were made in the control of the Concorde project. In 1968 the French Minister of Air Transport was replaced and his post was renamed Secretary of State for Civil Aviation. In July of that year the confusing interregnum of Maurice Papon as Chief Executive of Sud Aviation came to an end, and in his place the French made a brilliant appointment. They invited General Henri Ziegler to take his place. Ziegler was a very considerable figure. He had been Director General of Air France, and later also of Breguet until that firm was absorbed by Dassault. He was a Polytechnicien, a graduate of the Ecole Nationale Supérieure de L'Aéronautique, and a man of marked personal dynamism. If I may telescope events and put them into perspective, his position was consolidated not long afterwards by his being appointed President of Aérospatiale, which was in effect the merger of the French nationalized aerospace industry. It took in Sud Aviation, Nord Aviation and the French ballistic missile programme.

On the British side, Archibald Russell reached retiring age and was succeeded on the executive committee by George Gedge. That must not sound like a bald way of dismissing Russ from the action. It was a signal event, and has to be recorded. Russ in his humorous, emphatic, not to say eccentric, way had probably been the dominating figure in the whole project. For ten years he had held inflexibly to his own conception of the shape, size, weight and range of a supersonic air transport plane. The justification of all altercations he had had with the

French, of his highly independent attitude within BAC, his original and discursive methods and, if I may so phrase it, his obstinacy, was that he had been proved triumphantly right. Concorde, in production, conforms to his original conceptions at the end of the 1950s and is an astonishing record in a turbulent and argumentative industry. He was honoured with a knighthood on his retirement and it was richly deserved.

I also found myself involved in all these changes. I had been obliged for several years to concentrate on the commercial exploitation of the VC10 and the BAC One-Eleven. With the medium-haul One-Eleven we had come up with a winner. It had provided a substantial proportion of the Corporation's profitable income in the commercial aircraft field. However, Sir George Edwards asked me to slot myself into the Concorde operation in conjunction with Henri Ziegler. I count myself fortunate. We set up a rapport immediately and we worked in happy harness until his retirement five years later.

In July 1971, Henri Ziegler and I became the designated senior officials in each company to control the Concorde project. The existing intercompany organization and all its committees were disbanded by us. Ziegler and I became jointly responsible for the management of the project and decided to meet frequently, as we were already doing, to take decisions on all those matters which could not be resolved by the directors of the two companies. Two project directors were appointed under us, Pierre Gautiér in Aérospatiale and Mick Wilde in BAC. Louis Giusta and John Ferguson Smith were appointed as our deputies in case of unavoidable absence but, principally, they were to be responsible for the general management of the programme within their own companies. Five management subdivisions were appointed – technical, production, commercial, contracts and flight test, the latter under André Turcat and Brian Trubshaw. It was a long overdue streamlining of the administrative structure and it enabled us to give much greater cohesion to the joint development effort in its final stages.

Henri Ziegler was the fortunate inheritor of the firmer French commitment to Concorde. In Britain we had been subjected to the whims of changing governments, and to attacks in newspapers, articles and books by committed critics of the project who had adopted the arguments of the economists and the environmentalists. From all this the French had been relatively free, although the waves of criticism in Britain had necessarily affected them. The main competitors to Concorde came from the United States and Russia, and the main criticism from America. It was perhaps the American attitude that determined the French. The very fact that the Americans did not wish to see a supersonic transport develop in Europe only hardened French resolve to see the project through. This unified all levels of their public life, political, popular and in the unions. Concorde was in many ways an overt act of defiance to counter American influence, and I know that the French often felt that the British wavered because we were more influenced by American policy than they were.

Henri Ziegler has presented this theory to me many times but I am by no means sure that I accept it. He overlooks our subjection to frequent changes of party politics which the French were spared during the whole period of the project. He also discounts the presence of the British Treasury, which is a law unto itself, survives all changes of government and regards expensive prestige projects such as Concorde with grave suspicion. It is the same under a Conservative Chancellor as under a Labour one, and in spite of the violent fluctuations in Britain's economic situation over the last twenty years or so, the Treasury consistently thinks in the short term. What is not bringing a visible return has to be questioned. The Treasury, which controls the purse strings, does not seem capable of thinking in a big way about projects that are of long-term importance to the nation, whereas in France either its Finance Ministry does not think that way or, if it does, it does not appear to have the same grip on the political situation.

Ziegler came to his post with overwhelming credentials. He was able to question decisions in a manner which his immediate predecessor would have found difficult. He was an engineer, a former test pilot, and he knew as much about the business as anyone under his control. It caused some friction to start with, but he certainly got things moving.

What Henri Ziegler and I had inherited was a very impractical system of control. On the grounds of efficiency and economy, there should have been single development, a single final assembly line, a single flight test centre. When I took over the top job, I had a very close look at whether or not, even at that late stage, we could concentrate on one assembly line and one flight test centre. It would have been very difficult to do, as it would clearly have involved a good deal of redundancy at one end or the other. Common sense would have dictated that the final assembly and flight test should be done at Toulouse because they have both facilities in the same place. The Filton runway is not satisfactory for the flight testing of Concorde, which is why this had to be moved a little further northeast up to Fairford. If the decision had had to be taken to concentrate it in one place, it would have had to be at Toulouse. This was discussed, very much behind closed doors, because if it had got out there would have been the most frightful hue and cry. I did not even take it up with the French. We simply examined it internally, on my insistence.

It was too late in the day to change, and furthermore it would have been politically totally unacceptable. To have concentrated the entire operation at Toulouse, the economy-minded solution, would have meant that by the time the airplane entered airline service the British would have entirely lost the capability for final assembly of aircraft at Filton. The people involved are different from those working in machine shops and other essential plants and, once we took the work away from them, we would have lost them. We would have entirely lost our flight test capability because we have no other civil aircraft development coming up and we would have had to disband the team. There was a certain amount of BAC One-Eleven production flight test work going on, but this was simply not work of the same proportion. The next European project that came up would have found us incapable of bidding for it because we would have disbanded our capability.

This was the powerful and overwhelming argument for carrying on as we were. But in terms of economic management, there is no doubt what the solution should have been years before, and no doubt that the form of organization that Henri Ziegler and I inherited added enormously to the cost of Concorde. There can be no argument about that. The duplication of jigs and tools and flight test facilities and the other additional costs automatically built in when you are doing things on a collaborative basis involve an extra dimension. I would have to concede that building airplanes in a collaborative fashion can add up to thirty per cent of the costs, which in the case of Concorde represents £300m. On the military side this sort of organization can make sense. If you have two air forces committed to buying the airplanes you are producing, and you have a market for 600 instead of 300 planes, then the additional cost can be justified. But in the commercial field you are not automatically extending the market to any great extent.

It is only fair to say that both BAC and Sud Aviation learnt many useful lessons from their Concorde experience. When we got on to later projects that were to be conducted on the basis of collaboration, both of us applied the lessons we had learnt. In the late Sixties a considerable body of opinion had built up in France, Germany and Great Britain which considered that there was a world market for what has since become known as an 'Airbus', that is to say a wide-bodied, twin-engined, short- to medium-haul jet, capable of carrying anything from 200 to 300 passengers.

In the first instance, preliminary studies were carried out by BAC with Sud Aviation, with whom, because of our mutual Concorde experience, we were regarded as natural partners. Running parallel with our own studies were further investigations being carried out between Hawker Siddeley and the French firm of Breguet before it was merged with Dassault. As far as I recollect, both Hawker Siddeley and BAC received

(Left to right) General Henri Ziegler, Sir George Edwards, Louis Giusta, Sir Archibald Russell, Brian Trubshaw

Sir Stanley Hooker, Technical Director of Rolls-Royce and 'father' of the Olympus engine, with a first stage compressor section of an early version of the engine

Concorde flight deck and forward fuselage assemblies in production at the BAC Weybridge factory

A rear view of the Concorde nose and forward fuselage section during building at Weybridge

The third production Concorde rear fuselage and fin are joined up at Weybridge. After being equipped to an advanced standard, the completed section is shipped to the final assembly centre in either Britain or France

The full electrical harness for Concorde, including over one hundred miles of wiring, is prepared on this full size assembly rig before final installation in the aircraft

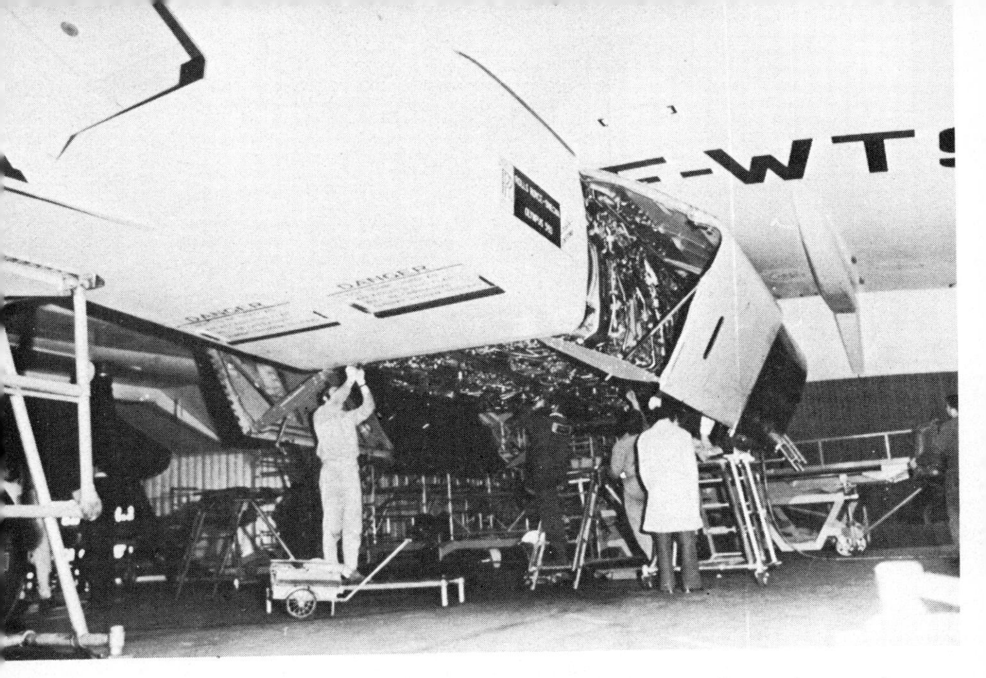

Concorde's massive twin underwing-mounted powerplant units, seen here on Concorde 02 F-WTSA, the French-assembled second pre-production aircraft

Concorde's famous test pilots André Turcat (left), Director of Flight Test of Aérospatiale, and Brian Trubshaw, Director of Flight Test of the BAC Commercial Aircraft Division, together on the flight deck of Concorde 002 at Toulouse

Concorde final assembly at Filton in the latter part of 1975. Series production aircraft 208 is in the foreground and 212 and 214 in the background

Roll-out of the first pre-production Concorde 01 (G-AXDN) from the cavernous 'Brabazon' assembly hall at Filton on 20 September 1971

The author (centre) being presented to the late President Pompidou of France at Aérospatiale, Toulouse, where the latter had just flown in the French-assembled prototype Concorde 001

Concorde 002 – the British-assembled second prototype and appropriately registered G-BSST – seen over Sydney, Australia, in June 1972 during its 45,000-mile sales demonstration tour of the Middle and Far East and Australia

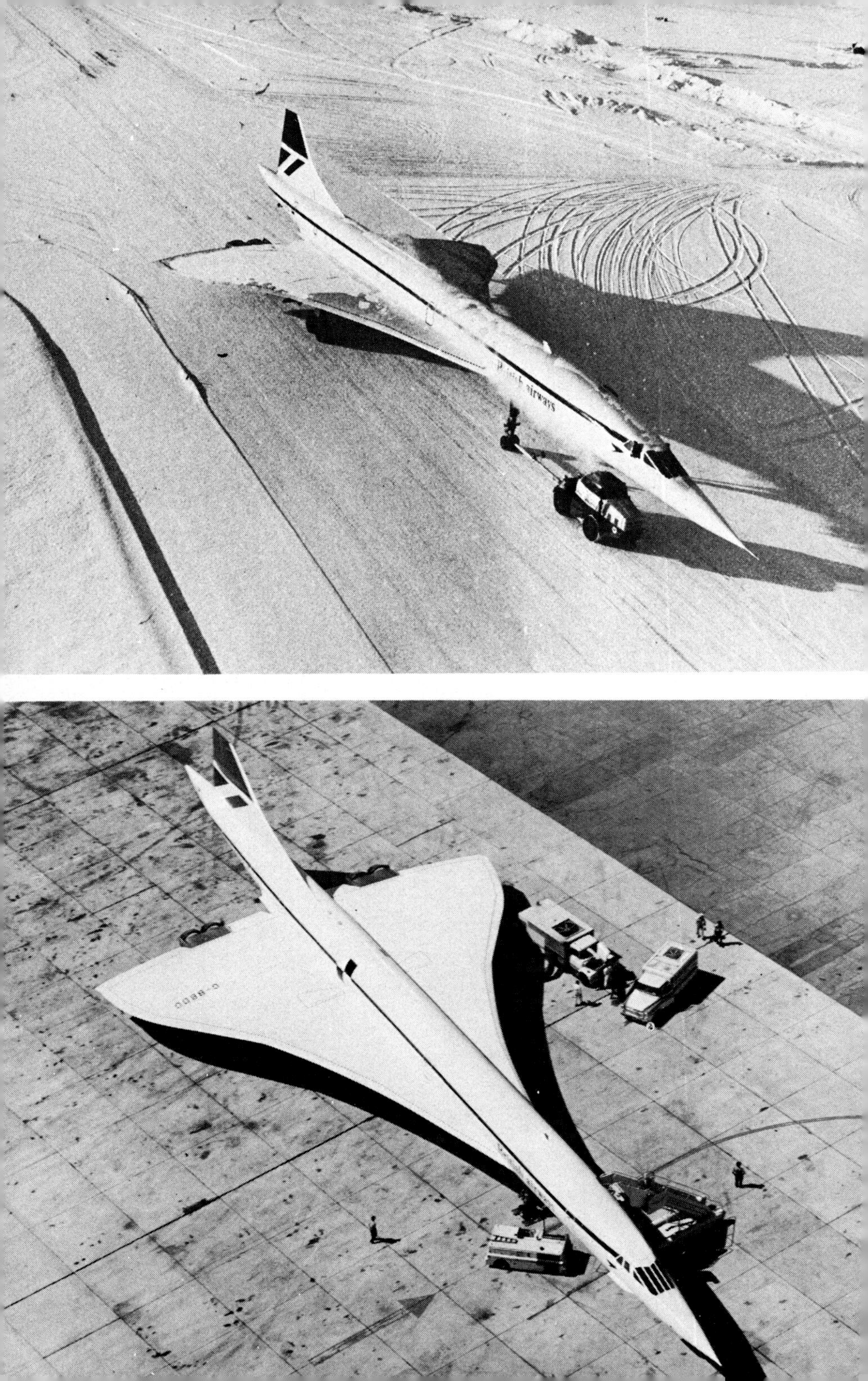

airways

Opposite above
Successful 'cold soak' trials in February 1974 with the (French-assembled) second pre-production Concorde 02 F-WTSA in the harsh and exacting environment (−44°C) of Fairbanks airport, Alaska

Opposite below
Concorde's clean lines and graceful shape seen from overhead – in this case G-BBDG, the second series production aircraft. Note the absence of tailplane, wing flaps and the many other protuberances common to subsonic aircraft

Above The author pictured with His Imperial Majesty The Shah of Iran on a demonstration flight from Tehran in Concorde 02 during its world tour in June 1972.

Below The Rt Hon Anthony Wedgwood Benn, Secretary of State for Industry and British Government Minister responsible for the Concorde programme, with BAC management and trades union officials at the time of a Concorde 202 demonstration flight from Fairford on 3 August 1974

Close-up of the production-standard Concorde engine exhaust units at the rear of the twin powerplants and of the aircraft wing

Opposite Concorde 204 – the fourth series production aircraft and registered G-BOAC – exhibiting the distinctive and characteristic ogival delta wing shape and underslung pairs of Olympus powerplants

G-BOAC

Concorde taking off. This head-on view further illustrates its distinctive shape, and the huge pairs of variable-area engine intakes are clearly visible

Concorde's famous needle nose (here in the fully up position but with the visor fairing ahead of the pilot's windows fully down) pictured beside its British stablemate, the BAC One-Eleven subsonic regional twin-jet

some small government financial aid for these early design studies.

Then the British government decided that it was only going to support one of these projects and on the basis, I must assume, of 'Buggins' turn next', chose to back Hawker Siddeley because they did not have a collaborative civil project, whereas BAC was involved in Concorde. In a further extraordinary twist, Hawker Siddeley were nominated to carry on studies with Sud Aviation, who were, in fact, BAC's partner, and we found ourselves closed out of this particular airbus project. In due course the Germans joined in with Hawker Siddeley and Sud Aviation, and it was agreed that before any decision was taken by the three companies to go ahead, certain criteria would have to be met. These included a demonstration that the three national flag carriers concerned were each willing to purchase substantial quantities of the new airplane.

While all this was going on, BAC, feeling that they had the necessary expertise for the job and confident that they were more capable than anyone else in Europe of actually marketing the airplane, came out with a development of their own wide-cabin, short- to medium-haul airplane of around 250 seats known as the BAC Three-Eleven, using the Rolls-Royce RB211 engine. BAC went out to get the orders and eventually obtained conditional agreements for nearly forty of them. However, we were unable to get government support for the BAC Three-Eleven, and the project had to be abandoned after we had spent several million pounds on the development.

The rival airbus, which became known as the A300, went ahead. What is significant is that the collaboration took the form of setting up a single company known as Airbus Industrie, based in Toulouse, with one flight test centre, one final assembly line and so on. Although the Germans are the main partners, with Hawker Siddeley involved to a limited extent, and also the Dutch and the Spanish, the whole direction of the project comes under one head from its French base.

This, I am sure, was a lesson learned from the much more

diffuse method of control employed in Concorde. In the case of the BAC Three-Eleven we had tried to set up a very similar organization. We had decided that while we would be very willing to have firms in Europe collaborating with us, it would have to be on the basis of single control from BAC at Weybridge. It is not without interest that if the BAC Three-Eleven had gone ahead, we were planning to bring in the Romanians and Yugoslavs as partners in the project.

The system we had worked out was this: we would allocate to a manufacturer, whether British or European, one component, for instance the wing. He would have the responsibility of carrying out the detailed design to our basic specification, designing and manufacturing the jigs and tools, and making the components for shipment to BAC. The supplier would finance his own design, tooling and manufacture, and would recover a percentage of his development costs with every component set he shipped to us, as we would with each aircraft we delivered. BAC was going to maintain the ultimate control of design, development, flight test and final assembly. So it was that, although our methods of operation were slightly different, both the British and the French, when it came to airbuses, set about organizing their construction in a very different manner from that employed for Concorde.

Henri Ziegler and I came to the Concorde programme with similar attitudes towards international collaborative projects. Moreover we both felt that the whole pattern of development was wrongly based. But we had to make the best of what we inherited. There could be no question of abandoning or watering down the project and it was absolutely essential to maintain the position in the market we had won. His reasoning was that if we dropped out of Concorde we dropped out of everything. He waxes extremely eloquent on this point, and considers that if the Europeans want to survive as a free and independent group of people they cannot let their guard down. He draws historical parallels and sees the European nations in the same situation as the Greek republics when the Roman Empire was building up. They had brought all the progress to the ancient world and were overthrown by the Roman Empire when they started fighting among themselves. For Henri Ziegler, Concorde is a major element in the European fight to keep its independence and freedom rather than succumb to American or Russian dominance. He feels there is a widespread tendency in Europe to abandon aerospace in the hope that America will protect us. He feels that this is dramatically wrong, and his drive to push Concorde through to airline operation has been based largely on this philosophy.

There is one substantial point on which he and I are not in full agreement. It is often said that one of the elements in the runaway cost of Concorde is that the manufacturers were spending government money rather than their own. We have enough experience in the British side of the industry to back the argument that this is not the case. We do not have two different standards, and we have always been as careful with taxpayers' money as we are with the funds of our shareholders. One effect of Concorde which I deplore though is that it has led to an increasing involvement of government officials in areas in which they are basically ignorant and in which they have no place. They have one major and legitimate function and that is to see that public money is being spent in a properly controlled manner – to use the British phrase, 'that there is full accountability'. With Concorde, government officials took unto themselves the right to second-guess our chief designers, sales and production managers. They demanded the right to determine whether the all-up weight should be increased by 5000 lb or not. I think that has been a great pity, and I regard it as wholly bad that government officials should be in this situation.

Henri Ziegler considers that the method of financial control has been a disaster. The controlled system of funding never set a given figure to which the industry was required to stick. He makes the point, and it is true, that the French government wanted the industrialists to agree to fixed prices and that Aérospatiale agreed and BAC did not. He considers that the behaviour of people in industry is entirely different if they are responsible for the cost of what they are doing and he does speak from enormous experience of the aircraft industry.

I rather tend to take a compromise position and consider that in the research and development period, where you are never quite sure what it is you are getting into until you are fairly well down the line, the work has to be paid for at cost under whatever system of control is necessary. Ziegler contends, however, that there comes a point when you have built your prototypes, when it would be a good idea to build the first batch of production airplanes for a fixed price. There would need to be an escalation clause to take account of inflation and other contingencies beyond a company's control, but this is something that we on the British side declined to accept, whereas on the French side they were prepared to do so.

However, as a further explanation of the differing attitudes towards this particular problem, it is perhaps worth emphasizing that whereas Aérospatiale is owned by the French government, BAC is privately owned and therefore tends to operate to somewhat different criteria. The French are just transferring money from one government pocket to another, whereas, in the last resort, BAC can only survive if it operates commercially and profitably for its shareholders.

Another view expressed by Henri Ziegler is that a multinational corporation is just the same as a national company. Once you assemble the financing capabilities of a number of people, using the corporation system, with a board of partners who discuss major policy, then they must delegate the power to a centralized management. There has been a history of successful private enterprise and the Concorde organization would have been the better for it had it been based upon that experience.

We have learned our lesson and in any future international collaborative projects will apply the results of our experience.

What you might call the Ziegler–Knight honeymoon was rudely interrupted in the spring of 1970 by the General Election in Britain and the advent of a Conservative government under Edward Heath. In BAC we drew a sharp breath because we knew from previous experience that we were liable to be pulled up by the roots again so that the way in which the Concorde operation was proceeding could be examined. The situation in the industry as a whole was not favourable. Rolls-Royce was already running into difficulties on its RB211 project with Lockheed, although this lay outside the immediate Concorde orbit. BAC was deeply committed to the Three-Eleven project. The Conservatives had come to power in a reaction against undue government participation in the economy and had appointed as Minister of Technology John Davies, a notable free enterpriser who acquired senior ministerial rank in one jump from being Director General of the CBI. He was well known to be opposed to lame ducks in industry and no enthusiast for unprofitable prestige projects. We feared a return to the bad days of 1964.

Our worries were temporarily alleviated by the appointment to the responsibility for aviation in the new enlarged ministry of Fred, later Sir Frederick, Corfield, who had been the MP for South Gloucestershire. His constituency happened to take in the Bristol factory at Filton, so several of us in BAC had known him for some time. We soon got a decision on the BAC Three-Eleven, which was not to go ahead. We knew that the problems of Rolls-Royce on the RB211 were accumulating, so that by the end of 1970 we were distinctly disconsolate. We simply did not know which way Ted Heath was going to jump on Concorde.

The Prime Minister took the very reasonable decision of requiring the 'Think Tank' he had organized under Lord Rothschild to carry out a fundamental study of the project. They certainly did their job thoroughly. Some of us assumed that the Civil Service would have tried to make this intrusive body into a fairly harmless outfit by appointing to it a bunch of ineffective people. Not so at all. The 'Think Tank' was an integral part of the Cabinet Office, and it attracted some very bright people. Cooper Brothers, probably the most high-powered firm of accountants in the country, were commissioned to look into the whole cost and estimating side of the project. Their senior partner, Henry Benson, was a most influential figure. There was hardly a major financial operation in which he had not been involved as an investigator.

Fred Corfield was not at all sure that he enjoyed having Lord Rothschild's 'Think Tank' and Cooper Brothers, the accountants, carrying out the investigations which he felt were more properly the province of his own officials. He went so far as to say that he disliked this method of government, and told me an anecdote that was apparently going the rounds in Westminster. Someone was looking for Henry Benson to give him yet another job, and he could not be found anywhere. Then it was reported that he was at Buckingham Palace, where the Queen had summoned him to form an alternative government. The joke did not make Fred Corfield smile.

BAC found itself subjected to the dispassionate and enthusiastic investigation of a small team of young, bright accountants and civil servants who came down to Filton and Weybridge, turning us upside down and delving deeply into the whole operation. In the end, and in quite a short period of time, they gave us a clean bill of health and recommended that the Concorde project should proceed. I have been told that when the report was placed before the Cabinet, Lord Rothschild, whom I got to know very well and for whom I have a high regard, had the pleasant idea of placing on top of each of the Cabinet Minister's folders a Concorde tie. Except that in the case of Mrs Thatcher, Aérospatiale had provided a Concorde headscarf made by Hermès. It was a nice touch.

Once the recommendation had been accepted, Ted Heath became a strong supporter of Concorde. He came down to Filton and went for a flight in the Concorde 002. We found him extremely easy to get on with. He was chatty over lunch and altogether quite different from the general public image that he has projected of himself. From that time on, until the Conservative government went out of office in the spring of 1974, we had his maximum support.

In the traumatic sequence of events that followed almost immediately after the crisis at Rolls-Royce, Lord Carrington found himself the responsible Minister, and in due course the aviation portfolio came under Michael Heseltine. He found himself in the fortunate position of administering a positive Cabinet policy to support Concorde. We found him a very thrusting Minister, who gave firm and decisive shape to the Cabinet directive. He took a prominent and most useful part in the first sales drive to sell the aircraft in the Middle and Far East and in the contractual negotiations with BOAC for the five Concordes they were eventually persuaded to order.

I am obliged to end this account of the situation in the early 1970s with some reference to the collapse of Rolls-Royce. It had less effect than might have been expected on Concorde, for the Rolls-Royce plant at Filton was building the engines. But this requires some explanation.

It is probably true to say that Rolls-Royce gambled the whole future of the company on developing the RB211 engine for the Lockheed 1011. Sir Denning Pearson, who was then the Chairman and Chief Executive of Rolls-Royce, had come to the conclusion that the only way the company could maintain its position in the world was by getting into the big aircraft engine league by developing a new-generation subsonic power unit with at least 40,000 lb thrust. He was determined to get the RB211 engine into a major American airframe development because, in his view, the only possibility for Rolls-Royce to continue in really major aero-engine production was to link themselves to one or more of the American airframe manufacturers, who had been so much more successful in selling their products round the world than had the Europeans. To do so, Rolls-Royce took a number of risks, both on the financial and also on the technical side, particularly in the use of carbon fibre turbine blades.

The terms of the contract with Lockheed meant that Rolls-Royce was heading inevitably for disaster. Most of us in the aircraft industry could see this coming, although few of us expected events to take the form that they did. It did not seem necessary to some of us to declare the whole Rolls-Royce empire insolvent. The government was going to have to bail them out anyway because of their defence work, and it would have been better, we thought, to deal with each component of the company on its merits. Rolls-Royce was, after all, something of a sacred cow, and the damage to the name involved in bankruptcy was shattering. It had been regarded around the world as the ultimate example of British excellence and integrity.

I suppose that in the aircraft industry we become impervious to disaster. Even when the crisis exploded, we never really expected it to have much impact on the Olympus engine for Concorde. We did not expect the liquidation of Rolls-Royce to mean that there would be no more engines from them. This would have affected the defence programmes of a number of other countries all round the world, which would have been thrown into total confusion if the company had just been allowed to disappear.

In retrospect I suppose it can be said that, although a lot of face was lost by letting Rolls-Royce go bankrupt, contractually they probably did fairly well out of it because they were able to renegotiate their prices for the RB211 engines and the whole deal with Lockheed was put on a more satisfactory basis.

In practice, the operation caused hardly a ripple as far as Concorde was concerned. It so happened that in BAC we owed Rolls-Royce rather more money overall for engines they had delivered than they owed us, so they were in a poor position to place us under any pressure. Almost as soon as they had gone into liquidation, an announcement was distributed around the world by Rupert Nicholson, who had been put in as liquidator, to airlines, air forces and manufacturers, to the effect that they would continue to look after their customers and supply engines according to contract. So immediate reassurance was given to everybody.

Our own operation went on as before. Geographically, Rolls-Royce at Bristol is on the other side of the runway from BAC. When they have an engine ready, all they have to do is trundle it across the concrete, and the arrangement continued as if nothing had happened. The site we occupy was originally set up more than sixty years ago as the British and Colonial Aeroplane Company. It is alleged that when Sir George White, the local magnate who founded it, set it up, he did so because he owned the local tramway company. This was the end of the line and it was a good way of improving traffic. I wonder what he would have made of the complications of industrial organization today?

9 Options and Opposition

Julian Amery had taken a very high-powered delegation of scientists and leaders of the aircraft industry to Moscow when he was in office in 1963. Their principal interest was in the Ilyushin IL62, which they found to be a copy of the VC10. Towards the end of the tour there was some discussion about supersonic aircraft, and the delegation, to its surprise at that time, was shown a model of a supersonic aircraft the Russians were considering. It was no more than a little model on the table, but they were absolutely staggered by its resemblance to Concorde. Rather a serious view was taken at the time, and there was a ministerial inquiry into the possibility of some leakage of information, but Archibald Russell always tended to pooh-pooh this, pointing out that once it has been decided to engage in supersonic flight the shape is more or less imposed on the plane for aerodynamic reasons. When therefore on 31 December 1968 the Russian TU144, generally nicknamed the 'Konkordski', flew for the first time, it came as no surprise to us.

The Russians had also arrived at an operational speed of Mach 2·2 and the similarities to Concorde were quite extraordinary. In their first version however the Russians had a great duct under the belly of the aircraft and four engines in the middle, which they have since modified. Archibald Russell gained the impression that it was a very heavy aircraft and that the Russians had been having considerable trouble with their engines. Like Olympus, these had originally been a military design, but we had good reason to doubt whether they had been sufficiently tested in sustained airline-type operation.

However it was not the Russians whom we feared as rivals but the Americans. From the earliest days of our supersonic aircraft studies the shadow of a possible American SST had overhung us. If we settled on a Mach 2, might they, in the course of their development, go for a Mach 3 and make a success of it? Would we end up with an obsolescent airplane almost as soon as it went into service? Would it be the repeat of the old story of the jet airplane coming in and making the turbo-prop obsolete?

During this initial period none of us had any doubt at all that in due course the great American aircraft industry on the West Coast would embark on a supersonic aircraft. President Kennedy had already appointed Eugene Black, a highly respected figure, to chair a committee to investigate the whole issue of a supersonic transport to be built by US manufacturers, and to report back to him. By this time Concorde was under way.

In the course of their investigations, Gene Black brought some of his team to look at the situation in Britain and France. I remember meeting him in his suite at Claridge's and suggesting that it would be a rather extravagant use of funds if both the United States on the one hand and Britain and France on the other were to develop competing models of supersonic aircraft. I proposed that the best solution might be for the Americans to join us in, if you like, a 'supporting' role on the Mach 2 airplane and that we in turn might join them in a supporting role on a Mach 3 airplane, thus ensuring an orderly development of the supersonic transport age.

In the end, Gene Black's committee recommended to President Kennedy that the United States should go ahead on its own in the development of a supersonic transport and that the major proportion of the funding should come from the US government. The recommendation was that ninety per cent of the cost should come from public funds and ten per cent from the firms selected. President Kennedy accepted the report, and two firms emerged from the first round of design studies – Lockheed and Boeing. Both had spent a fair amount of money in putting forward their proposals for Mach 3 airplanes. The Lockheed one had a double delta wing shape. The Boeing design was much more sophisticated. It was a swing-wing airplane which, from the design aspect, had the advantage of obtaining optimum performance out of the airplane in one configuration subsonically and when swept back gave the best performance in supersonic flight. This was the design that impressed everyone, and Boeing were eventually given the development contract.

A lot of people in Britain, including Russ, believed that Boeing were being over-ambitious, although they were a highly experienced and successful company. We felt that attempting to design a big Mach 3 supersonic commercial transport and at the same time to solve the problems of the variable sweep or swing-wing was too much. Our predictions were correct. After Boeing had spent a considerable amount of money on development, they decided that their first design was not satisfactory. In essence they tore it up and started again, ending up with the development of a delta wing airplane not very dissimilar in general outline from the one that Lockheed had put forward, but with a separate tailplane.

By now we reckoned that they were about five or six years behind Concorde. What is even more interesting is that as their design progressed, it turned into an airplane only marginally faster than Concorde, with a Mach number of about 2·7 compared with Concorde's then 2·2. From the speed point of view there was going to be very little difference in operation, although the Boeing airplane was going to be a good deal larger.

The funds for the Boeing development had to be approved each year by Congress, and by 1970 it had become clear that a considerable body of opposition had built up to the continued funding of the SST. The opponents in Congress and the Senate were something of an unholy alliance of people with two different objectives. On the one hand there were the environmentalists, who increasingly at that period in the US were taking a stand against advanced technology on the ground that up until that time technology had been the god, and the other important things in life had tended to be swept on one side in order to support its advancement. The second body of opinion was led by the Democratic Senator William Proxmire of Wisconsin. It was violently opposed to government intervention in private industry. His view was quite simple. If the American SST was worth doing at all, then it was worth while for Boeing to fund it. If Boeing was not prepared to fund it then it was not worth doing.

This opposition made great headway in 1970, and the project was very nearly cancelled in that year. Matters started coming to a head again in early 1971, and at BAC we found ourselves in the position of supporting Boeing in their fight. Our reasoning was perfectly logical. We felt that if the American SST was cancelled, then life could very well start to become much more difficult for Concorde. We would have to face up to the certainty of powerful forces in the USA determined that we should not succeed in a contest from which they had eliminated themselves.

Then on 24 March 1971 the American Senate voted against continuing financial support for the supersonic transport being developed by Boeing as airframe contractors and General Electric as engine builders. By the time the project was cancelled the Americans had spent more than $1000m, with virtually nothing to show for their money except 13,000 pieces of hardware parts, tools and raw materials and 2000 boxes of engineering data, records and correspondence gathering dust in a warehouse. Nothing in the shape of a complete airplane, and the loss of more than 7000 jobs.

However Pan American Airways were not caught entirely unawares. During the initial stages of the development of Concorde, in the early 1960s, they had hedged their bets. Juan Trippe, founder and at that time still President of Pan American, sent over to Europe a very old and trusted colleague, Franklin Gledhill, whom I had known for some years. His main declared interest when we met was in a Hawker Siddeley business jet called the HS125. Gledhill had come to Britain to see if he could do a deal with Hawker Siddeley which would enable a new Pan American subsidiary in the United States to act as distributors for the aircraft, and this would of course have given it a major place in the American market.

At the same time he apparently had reserve instructions to see if he could take some options on Concorde, and he invited Allen Greenwood, now the chairman of BAC, and myself, to have lunch with him at Claridge's. There was, I think, a dual motivation behind his request. An option agreement on Concorde would provide a spur to the American SST project, which had not at that time been given the full go-ahead. If in the end no American SST emerged, then Pan American would be protected if BOAC and Air France got on the North Atlantic with an airplane which would otherwise have threatened their business.

But in 1963 BAC had no wish to negotiate options for Concorde. We were in the early stages of the programme, and an option to purchase would inevitably have had to contain a number of clauses enabling the option-holder to get out of his commitment relatively easily. Sir George Edwards was very much opposed to option deals. We were under no particular pressure at the time to demonstrate that we had a saleable commodity, and we all decided that we would rather wait until we were in a position to invite firm orders. It is fair to add that the French were not necessarily of the same mind.

Frank Gledhill persevered, and clearly had Juan Trippe in New York telling him over the telephone that he was not to give up. His negotiations with Hawker Siddeley were not progressing very well. I think Pan American wanted to substitute the British engine in the HS125 with an American powerplant. Nothing came of the deal, and Frank Gledhill adjourned to Paris to start talks with Marcel Dassault to see if he could do a similar deal for a business jet derived from the Mystère fighter. In this Gledhill was successful and the American version was subsequently renamed the 'Fanjet Falcon'.

All these transactions took place during the period preceding the biennial Paris air show, an event most of the top people in our business attended. We found ourselves in the company of Sud Aviation, talking again to Frank Gledhill, and it became clear to us that he had made a good deal more progress with the French about Concorde options than he had with us. To cut a long story short, BAC changed its attitude and with Sud Aviation granted Pan American an option on six airplanes, three from each of us, which involved a quite handsome payment of £210,000 to BAC and of nearly 3m francs to Sud Aviation.

I remember the occasion particularly vividly because the Paris manager for Pan American called round at the office in Sud Aviation of Paul Simonet, who had been my partner in the negotiations, to deposit the two cheques. I went round with ours to Lloyd's Bank in Paris to pay it in to our London account. There was something of a queue, and the counter clerk was distinctly offhand with me until I waved it in front of him when he nearly fell over backwards at the amount. I was pleased to get rid of it as I did not particularly fancy walking round Paris with a cheque for half a million dollars.

The Americans are good negotiators. Juan Trippe had insisted on a clause very favourable to him: he was to get one in every three of the production airplanes up to the first eighteen. This meant that if and when BOAC and Air France put Concorde into service, Pan American would be in a position to compete with them and not run the risk of having all their first-class traffic taken away by the British and French.

What happened next was almost ludicrous. The Pan American options triggered off something like a minor stampede from the major airlines of the world. At the same time the Federal Aviation Administration in the United States started selling options on the American SST, although at that time there was not only no such airplane in the development phase, but no decision had been made between the rival proposals of Boeing and Lockheed. A queue formed in America for an airplane whose existence was by no means certain and whose size, shape, capacity and performance was a totally unknown quantity. It really was one of the strangest competitions in aircraft history.

In the end the FAA obtained 122 options from twenty-six airlines. Deposits had to be paid, but they could be returned on demand from the airlines, so that this was rather a different situation from the tighter contractual arrangements that the British and French were making for the Concorde option list. Nevertheless, this was regarded as an international competition by the newspapers, and great play was made of the fact that Concorde only attracted seventy-four options in comparison with the American figure. All this tended to prove that Sir George Edwards, who has an uncanny prescience about this sort of thing, was absolutely right in his unwillingness to enter into option arrangements in the first place. As subsequent events proved, we would have avoided an awful lot of difficulty and trouble if we had said no to Juan Trippe and meant it.

There were of course some marginal advantages about these option arrangements. All the airlines which had taken them out were required to pay us a deposit, and for a long time we were able to sit on this money without paying any interest on it, although we did so at a later stage. The second option after the Pan American deal went to an old American friend of mine named Bob Six, the extremely colourful President of Continental Airlines. He had decided that if the supersonic age was about to begin then he wanted to be out in front. He negotiated options for three Concordes with me, although not on the same favourable delivery terms as Juan Trippe had obtained. After that the list started to grow rapidly – Continental, American, TWA, Eastern, United, Braniff and Air Canada, all in North America, and Middle East Airlines, Qantas, Air India, Japan Airlines, Sabena and Lufthansa in the rest of the world.

In the meantime we were falling behind schedule with Concorde, and talking in terms of delivery dates a good deal later than those we had put forward in our option agreements.

Nevertheless, we received very few complaints from the airlines on this score, and the reason was quite simple. We were now entering a period when the major airlines of the world were busy ordering and putting into service the Boeing 747, first of the jumbo, wide-bodied subsonic jets. Pan American had started this process by ordering twenty-five of them in 1966. Then the race was on around the world to put jumbos into service from 1970 onwards. The airlines were so much involved in this operation that they were quite content to see Concorde slip back in time. Indeed for most of them Concorde was a future worry which they would just as soon have postponed as long as possible.

During this quiescent period I was appointed to the board of the Weybridge Division of BAC at the beginning of 1965, became its Managing Director in 1966, and Chairman in October 1967. My main occupation there was to sell the BAC One-Eleven. I became, in addition, Chairman of the Filton Division in October 1969, which brought me back into the Concorde field, and then in June 1971 I became Chairman of the newly formed BAC Commercial Aircraft Division, which took in both Divisions.

It was only three months after I had taken over the responsibilities at Filton that another blow fell. *Time* Magazine had seen fit to organize an extremely high-powered seminar on the subject of Concorde. It flew over a chartered planeload of leading American airline chiefs and businessmen, invited a whole covey of their equivalents from several European countries, and brought them first to Filton. We showed them round, put them up at the Grand Hotel, and held a one-day conference at which Anthony Wedgwood Benn made the main pro-Concorde speech. That was on 6 February 1970. The next day the whole circus moved on to Paris. One of the American delegates was Najeeb Halaby, who had succeeded Juan Trippe as President of Pan American.

The main conference in Paris was held in the new Air France centre on the left bank. General Ziegler was on the platform, so was I and so was 'Jeeb' Halaby, as he was always called.

He is an interesting character, ex-lawyer, ex-pilot, ex-administrator of the FAA before Juan Trippe brought him into Pan American and built him up as his successor. He is probably one of the most articulate Americans I have ever met. He is an extremely good and entertaining speaker and can hold an audience off the cuff. On this occasion, he spoke from a written text, which was then circulated.

He started off in genial enough form. 'I think we really should congratulate this fabulous group of gallant Gauls and brave Britishers for undertaking and bringing this great project along this far, so well, so soon,' he said. 'There is no question but that it is a technological tour de force. It is one of the most fabulous examples of forced technical growth in the history of technology and of man, really.'

So far, so good. He then went on to extol the virtues of the Boeing 747 in which they were making the trip, and then started making unfavourable comparisons with Concorde. In a reference to Tony Benn's speech the previous day in Bristol, he went on: 'I am afraid that in the physical dimensions of Concorde, contrary to what the Minister was suggesting yesterday, the public will not consider the 747 airbus an old-fashioned antique airplane. They might well consider the dimensions of the interior of Concorde an old-fashioned ride because of its construction, back to the tube, you might say, from the living room.'

His main theme was that Concorde would have to be regarded as a successful experimental plane which should make way after a few production models for something bigger and better. 'Are you prepared to produce a very limited quantity of Concorde I, and then immediately face up to planning and financing Super-Concorde?' he asked. 'An airplane which will better meet the requirements of both passengers and airlines and will be superior to the Soviet SST and competitive with the US SST? ... We think that engine definition should proceed now for Super-Concorde, and that it ought to be completed by the time suggested for the completion of engine definition for Concorde I in 1973.'

I was incensed by this, but at the same time stunned because the other people on the platform connected with the Concorde operation hardly reacted at all. Henri Ziegler, who is a pretty emotional character, seemed to be taking it very calmly, and the rest of them seemed to regard this speech as an academic appraisal of the project which did not call for immediate comment. There was a brief question and answer session after the speech which meandered on. Then when I got to my feet I counted ten before I let fly. My exact words, which I remember to this day, were: 'I feel a hot trickle down my back where I think Jeeb Halaby's knife went in,' to which he interjected – as I have said he is a very articulate, quick man: 'It wasn't a knife in your back, Geoffrey, it was a needle in your backside.'

I have always felt this was the first step in a sequence of events that could have destroyed the Concorde project. The American SST was still very much in the running at the time and American opposition to Concorde has been so consistent over the years – witness Julian Amery's recollections from the early Sixties – that I gained the impression they felt something had to be done before it became an operational reality.

I think that there was also another factor which only emerged somewhat later. Pan American must have been starting to run into very serious financial difficulties. This was not yet apparent from their public accounts, but the downturn they and other American airlines were about to experience was already on the way. It would be understandable if they simply did not want to be faced in that period with the unpleasant decision of whether to continue their option agreements or not. It was a very subtle approach. Due acknowledgement was made of the technological achievement of Concorde. The fate of the American SST had not yet been decided and, if Concorde could have been dismissed to the limbo of a worthy experiment, it would have meant that the American version could have obtained a new lease of life and only become available for airline operation when a recovery in their own financial position enabled them to purchase it.

The next crunch came shortly after BOAC and Air France

signed firm contracts with BAC in London and with Aérospatiale in Paris at the end of July 1972. In the first instance, BOAC were to take five aircraft and Air France four. This meant that the clock had started ticking on the first of the option agreements, that with Pan American, because they had agreed to come to a final decision within six months of orders being placed by the British and French national airlines. So right through the autumn we found ourselves involved in another round of negotiations with Pan American. There was a lot of discussion but no feeling that we were really getting to grips with the situation. By this time Halaby had left Pan American and had been succeeded as Chairman by Bill Seawell. However the atmosphere had not changed, except that it had become more diffuse and less concrete. We found ourselves involved with a number of Pan American Vice Presidents from whom we could get no clear answer, although we gradually gained the impression that in the end it was likely to be negative.

This was no longer a straight commercial negotiation, as the British and French governments had become involved and, although we were required to answer a whole lot of questions from Pan American, we were accumulating them in a package for answer so that the two governments who were looking over our shoulders would have the right to take a final decision before any contract was signed. About the second week in January 1973, Sir George Edwards came over to New York to join me and I put him in the immediate picture. My recommendation was that we should place the onus on Pan American to 'fish or cut bait', as the American phrase has it. I got on the telephone to our contracts manager and our sales engineering manager back in Filton, and told them to get on the next plane and come to New York for a week to put together a package answer to all the American queries that they could either accept or reject.

I still hoped that Pan American would ask for an extension of their option beyond the end of January, as this would have cost them nothing and would have given us a further opportunity to persuade them to take Concorde. But you develop a feel for these things and I had the very strong impression that we were not going to get such an extension. The main reason why I wanted one is that if you are losing a competition, the thing to do is to keep it open as long as possible. A quick answer is inevitably going to be the wrong answer if you are on the losing end. If you are winning a competition, then the sooner you can close the deal the better. However Pan American were not behaving like people who were about to buy an airplane, and I have been in the business long enough to be able to sense that situation when it happens.

The other thing you discover in all these negotiations is that when someone is not about to buy, he thinks of every reason for avoiding a meeting, and there seemed to be a lot of that going on. Sir George Edwards eventually went back to London before the end of the month when the decision was due, and right at the end of the month Bill Seawell had to go to Geneva for a most important IATA meeting. I therefore found myself during the last critical days dealing with his immediate deputy, Bill Crilly. We met on 30 January, the day before the option deadline, and he undertook to give me a final answer the following day. What we did agree was that there should be a joint announcement and Press release, and on this Pan American were going to call in a senior colleague named Willis Player, whose title was Senior Vice President for External Affairs. He was, in fact, their top public relations man.

We met again on 31 January, and Bill Crilly told me that their decision was not to continue with the option. He then stuck in my hand a Press release prepared by Willis Player. I bridled at this and pointed out that the announcement was to be an agreed joint version. Player's draft certainly put the record straight as far as Pan American was concerned, but it was damning from the Concorde point of view. I had to point out that there were ways in which we could retaliate if we issued a separate statement, particularly if we referred to their inability to buy any airplane at all, never mind Concorde, in view of their financial position. I did obtain some minor modifications in the Player draft, but it did not make pleasant reading. This is what it said:

> Pan American will not exercise its options to purchase Concorde. Pan Am's studies indicate that the airplane will be capable of scheduled supersonic service but, since it has significantly less range, less payload and higher operating costs than are provided by the current and prospective wide-bodied jets, it will require substantially higher fares than today's. Concorde does not appear to be an airliner that satisfies Pan Am's future objectives and future requirements as the company now sees them. However, Pan Am will maintain an 'open door' to the manufacturers of Concorde for any new proposal they may wish to make.

The reader will appreciate that the last sentence of that Press release was not in Mr Player's original text.

The next day, three separate and more or less simultaneous Press conferences were held, one by Sir George Edwards in London, one by Henri Ziegler in Paris and one by me at the Pierre Hotel in New York, at which we explained what had happened. Almost the only ray of humour that emerged from the whole affair came from the representative of *Paris Match*, who attended my Press conference in New York. The whole first page of his report hardly referred to Concorde or the state of play with Pan American, but concerned itself with what I looked like, describing me as '*comme Prince Philip lui-même*'. On the full page opposite there was a picture alleged to be of Mr Geoffrey Knight of the British Aircraft Corporation but which was, in fact, a photograph of Sir George Edwards.

There were those who regarded the Pan Am withdrawal as the death of Concorde. I had been through this sort of thing before on a number of projects, and had become fairly wise in the ways of the aircraft business. When you are dealing with expensive capital goods, you are fortunate to win ten per cent of the battles. The only part I had found unpleasant was the business of the Press release. The Americans are pretty quick on their feet in these situations, and Willis Player was merely acting in what he considered to be the best interests of the company that employed him. But it certainly formed no part of the agreement that I thought I had with their management.

Pan American got their money back, as did most of the other airlines who were on the option list. There were three exceptions. Air India, Japan Airlines and Lufthansa maintained the options they had entered into.

The incident caused a considerable furore at the time, although it has to be said that the Press accounts which suggested that there had been a violent internal row between Pan American and BAC were wide of the mark. These negotiations are always conducted in a very civilized fashion, and nobody had raised his voice at any stage. You are dealing with such valuable properties, involving such huge sums of money, that people do not throw tantrums. People are far more likely to lose their tempers if they feel they are being diddled out of a tenner on some deal involving a second-hand car. The plain fact is that the options had never been determining documents and there was no point in placing more value on them than they had acquired at their first signature ten years earlier. The battle for the prospects of Concorde as an operational aircraft still lay in the future, and it is a story which will not even be complete when this book is published.

Perhaps the most heartening result was the closing of ranks in Britain and France. On the very day that the cancellation of the Pan American options was announced, the BAC combined committee issued a statement saying that: 'All Trade Union organizations throughout the British Aircraft Corporation reaffirmed their complete faith in the future of Concorde.' The news release put out after Sir George Edwards' Press conference in London, at which he expressed his own confidence and faith in the future of Concorde, really brought everyone together. Similar statements of confidence were made by Mr Michael Heseltine, Minister for Aerospace, Mr Wedgwood Benn, Opposition spokesman on Aerospace, and by BOAC, who described Concorde as a 'damned fine airplane'. The President of Air France said they would be operating their fleet of Concordes at fare levels ten per cent below first-class fare. The French Minister of Transport and the President of Aérospatiale both reaffirmed their belief in the success of the project, while M. Cot, Managing Director of Air France, said that, after 1975, there would be two kinds of airlines – 'those with Concorde and those without'.

We had received a serious setback, but the project went on.

10 Inflation

While we were going through these awkward experiences with the Americans, the Concorde planes themselves had been quietly marking up impressive milestones. President Pompidou of France had become the first Head of State to fly supersonic when he was taken up in the French 001 in May 1971. In December of that year he used the plane to fly to the Azores to meet President Nixon. In September it had made an absolutely trouble-free fifteen-day tour of South America. In January 1972 the Duke of Edinburgh had piloted the British 002 during a two-hour supersonic flight and then, in February, Concorde 01, the British pre-production model, flew supersonic for the first time. In June, we took 002 on a 45,000-mile sales demonstration tour of twelve countries in the Middle and Far East and Australia, which it accomplished without a hitch. July and August had brought us great encouragement with the firm contract from BOAC to buy five of BAC's Concordes, with Air France undertaking to purchase four from Aérospatiale. At the same time, the People's Republic of China signed a preliminary purchase agreement for three Concordes.

The original Franco-British agreement in 1962 had called for the two firms to set up an integrated sales organization. Neither BAC nor the French saw the necessity of implementing this initially. In the early stages, until we agreed on a common design, the two companies were offering a different aircraft, the French medium-range overland and the British long-range transatlantic. When the design was integrated we made no change and as we coordinated the Concorde operation very closely, the necessity for a single sales organization never arose and no problems were caused.

The two teams met regularly, every couple of weeks or so, either in Paris or London, and they worked in practice as a single organization. There was a period when we thought of splitting up the world market on a regional basis, but it did not really make any sense to say that Australia was a British sphere of influence and South America was French, for example. Both companies had particularly good contacts with some airlines and less close contacts with others, and it seemed better to allow the personal approach to predominate. It was very rare for either BAC or Aérospatiale to take the leading role in any particular sales activity without having some people from the other company on the team. We were going to share the proceeds equally anyway, so whoever got the order benefited both.

The French were for some time much more active in the promotion of Concorde than BAC was. This was particularly true after Henri Ziegler took over as Aérospatiale's President. He was ideally suited to the challenge and was a man of great zest and energy whom I account myself very fortunate to have had as a partner. He had a breadth of background which would be very difficult for anyone in Britain, with our more rigidly compartmented society, to match. He had a great deal of influence during the Pompidou presidency with top politicians, most of whom he knew well. He had been Chef de Cabinet to Chaban Delmas for a time, and had done his stint in government service. He was a technical general and an airman who piloted his own plane. He had run Air France as Director General, had entered private industry to run Breguet before it was taken over by Dassault, and had then gone to Sud Aviation which in 1970 he merged with other companies to form Aérospatiale. Top people in France move like this all the time, from government departments to public and private industry and back again in a way which does not happen on the same scale in Britain. He was very much a product of the French system.

Ziegler was a great promoter of Concorde and brought enormous enthusiasm to the task. He was very anxious to get out and sell the airplane, and it was he who organized, against a certain amount of opposition, the first series of demonstration flights which took 001 to South America. BAC was distinctly slow and reluctant to follow him in this, and when we took 002 to Tokyo and Sydney in June 1972, a number of our people had serious misgivings. They thought it was taking too much of a risk with a development airplane: not because it was unsafe, but because in the event of things going wrong this could mean that, instead of adhering to the programme, we might find ourselves stuck on the ground in Singapore or Darwin for a week waiting for a spare part from Europe. The world's Press might then say: 'There you are, we told you so, the plane is no good for airline service.' However, the early models of Concorde turned out to be incredibly reliable when they were taken away from base. We were able to do exactly what we said we would do when we embarked on these arduous tours without the full back-up facilities that would be normal in full-scale airline operation.

I think it is probably true to say that BAC tended to supply the larger proportion of airline marketing expertise. First of all we had a great deal of experience in this area, and secondly Aérospatiale, whose sales organization was smaller than ours, had to shift over a number of their people on to the A300 airbus project, which left them a little thin on the ground as far as Concorde was concerned. There had been some very substantial changes in the senior management of Aérospatiale, with a number of new people coming in from the missile business or the military side, and I would say that BAC was able to make up for their relative lack of experience in the commercial aircraft end of the business, although perhaps the French would not see matters in quite the same light.

There has been a certain amount of ill-natured comment in the Press about the French trying to upstage us on these promotion tours, but the accusation really does not hold water. The French were always more ready to take one of the models out of the development programme and show it to people. This is due as much as anything to a basic difference between the French and British character and outlook. The whole British philosophy, both with officials in government departments and with BAC, was to get on with the job of developing the airplane and getting it into service as quickly as possible. Sales tours can be an irrelevancy because an airline is not going to buy Concorde just because it looks pretty when it arrives on an airfield. We had organized a tremendously complicated and lengthy pattern of flight and technical tests, which we preferred to adhere to, so the French models seemed to be more frequently available for showing the flag than ours.

The feeling, primarily in the British Press, that the French were constantly engaged in one-upmanship probably had its source in the very first flight of 001 from Toulouse, which predated ours by five weeks. The French, quite rightly, made a great day of it and had all their bands out, but it had been a matter of firm agreement between the two companies that 001 should have priority in the supply of components as they became available. It must not be forgotten that the French were building sixty per cent of the airframe and BAC forty per cent. The proportion for the engines was the other way around, but then engines do not fly on their own. From that day on, the British Press was always looking for a British first, but somehow it never happened.

Although the French made their models available for a number of demonstration flights more freely than we did, it was largely fortuitous, and the arrangements on every single occasion have always been meticulous in the sense that the French have always insisted on us being properly represented. The aircraft have had the Air France livery painted on one side and the British Airways livery on the other. They have never tried to high-hat us on any of these trips, but the impression that they did is still there.

Once the Heath government had made up its mind to back Concorde, we received a very generous measure of ministerial support. Michael Heseltine had become the Aerospace

Minister, and he very much put his reputation on the line as far as Concorde was concerned. He went on most of our Far Eastern tour and had the satisfaction at our stop in Teheran of hearing the Shah announce that he was going to buy. In Bahrain, both he and I were given ornamental daggers to mark the visit. Then at Singapore, Heseltine and his wife returned to Britain and Lord Jellicoe, who was then Lord Privy Seal, took over the ministerial representation. I returned from Tokyo and Sir George Edwards joined the tour in Australia. It was an extremely successful and worthwhile trip. Japan Air Lines confirmed their option with an amendment taking it up to the end of December 1975, and later in the year Iran Air signed a preliminary purchase agreement for two Concordes with an option on a third.

Henri Ziegler and I spent four years trying to streamline the organizations we had inherited. We did think up a variety of solutions, including that of forming a joint company, but decided it was too late in the day and would accomplish little. Instead we decided to concentrate on practicalities. Regular meetings with masses of people sitting round a table did not seem to achieve much so we laid down that the project directors on the French and the British sides would continue in the fashion to which they had become accustomed, taking decisions on matters that needed to be dealt with on a day-to-day basis. As and when they got to the point where either they, or people from outside, or Ziegler and I considered that we had a collection of problems that needed a decision at the top, he and I would hold very short meetings to take decisions. We often met just *à deux*, or with no more than one or two other people with us, and then made swift work of clearing the whole agenda. I would like to think that this worked pretty well.

I do not want to make it sound as if Henri Ziegler and I rescued the Anglo-French administration of Concorde from years of confusion. A great deal of sensible rationalization had taken place continuously over the whole period. The original cumbersome Committee of Directors set up in 1962, on which I had had my baptism, first met once a quarter and after three years or so hardly met at all, for by then Archibald Russell and Louis Giusta had taken over at the practical level, although they were always the principal servants of their own companies. At that level of management and below there were continuous meetings and contacts. Where we had something of a total vacuum was right at the top, and we had acted to correct that situation.

We were still saddled with the two government committees of civil servants, and since they were our paymasters we had to maintain constant contact. They met fairly regularly, about once every three months, with a good deal of formality. The firms were normally asked to sit in for at least a part of each of their meetings and they tended to take decisions on virtually every subject which involved the expenditure of money. Most decisions involved this, so the parallel administration maintained its full vigour.

The quadruplication of correspondence was formidable. The opposite numbers in each company kept each other very well informed indeed. The method of communication between the governments and the companies took the form of the Chairman of the Concorde Management Board – a post which alternated annually between the French and the English – writing to the chairman of the company project department then sending a copy of the letter to the other side. So the directive would go out in French and English to both companies from whoever at the time was chairman of the officials' committee. The companies in turn would agree on replies which would be identical, one in French and one in English, to the official bodies on both sides. Anything that went from the French officials to Aérospatiale would automatically come to us, and vice versa. Aside from that, there would be daily discussions between BAC staff and members of the staff of the Director General, Concorde, in the appropriate department in London. The same thing went on in France.

Strangely enough, in spite of its volume, communications were relatively efficient. If the chairman for the time being was a Frenchman, the letter would go out to both companies in

French and we would then prepare a translation. But the communication itself would have been agreed between the two government departments before it went out, and in the same way our reply would be agreed between the two companies before it went in. Although everybody was reasonably well informed about what was going on, the system did have one weakness. Everything took at least four times longer than a single exchange of correspondence between a company and a ministerial department.

Ziegler and I used to meet ad hoc either in London or Paris, whichever was more convenient. He was always ready to come over to London. He had his finger in every pie and was always enormously concerned about what the Press were saying about the project. This caused considerable embarrassment from time to time because the British Press over the years has come out with some really dreadful stories about Concorde, whereas by and large the French Press has supported the project, with the single exception of Jean-Jacques Servan-Schreiber.

I never went into Ziegler's office in Paris without finding him with a collection of press cuttings from the British papers on his desk. They used to upset him badly when the comment was adverse. Sometimes I wondered whether he ever understood that we had no way of controlling it. I don't think he realized that in many cases this adverse comment had only a marginal and temporary effect on people's thinking. We were hardened to this, and had learned that an article that caused a furore for twenty-four hours was totally forgotten by the end of the week. He never seemed to accept this, and was liable to sense a sinister conspiracy behind it all.

At the end of 1972 I became Vice-Chairman of BAC as a whole, and my responsibility for the Concorde project became advisory rather than direct. Ziegler retired as President of Aérospatiale at the end of 1973 and so the Knight/Ziegler partnership broke up. He was a marvellous man, and I had greatly enjoyed my years with him.

Ziegler was succeeded by Charles Cristofini, whose whole

style of operations was quite different. Nothing like so fluent in his English as Ziegler, he preferred to have his meetings in Paris.

But this is to jump ahead a little, although we are getting near the present day. From this 1972–3 period, I do need to record one further development in the Concorde story – the advent of runaway inflation. At the end of 1972 there had been a major debate in the House of Commons during which it was established that the research and development costs had reached the figure of £1000m, half of which fell on the United Kingdom. The government was seeking authority for production finance, which it obtained, increasing the bill by a further £250m, again, as always, to be shared half and half. This brought the cost of Concorde, less the revenue in sales, to eight times what had been first estimated in 1962.

This escalating cost has caused much adverse comment. I have been frank about the causes of delay that lay with the companies: the serious differences in the initial conception; the long-drawn-out change in specifications to meet the requirements of the airline companies and transatlantic flight; the cumbersome nature of the international administration which was set up to control the project; the duplicated facilities for building aircraft and engines and testing them. In my judgement these delays must have added at least thirty per cent to the overall cost, but no one, until it happened, was able to take into account inflation of the magnitude we have experienced in Britain in the Seventies. In the middle Fifties we were paying a great many of our people down at Filton £10 a week. That became £40 a week and is now £80 and more. No commercial firm can forecast an economic revolution of this magnitude.

The old happy rule of thumb of selling an aircraft for £1 sterling for each pound of its weight disappeared shortly after the war. The initial estimate of the cost of a Concorde to the airlines was $11m, the figure in the option agreements which had been cancelled. As our next crisis approached, the figure had gone up to well over $25m. As Britain came up to another change of government in 1974, the auguries were not good.

But before moving on into 1974, there is perhaps one interesting event in the supersonic field which occurred in 1973 which I should mention: the crash of the Russian TU144 'Konkordski' at the Paris Air Show. I did not actually witness the accident in person, but I have looked at all the film of the crash and discussed it at length with our experts. None of us really knows what happened, and no report has ever been published, but there is little doubt that the Russian pilot got involved in too steep a climb. There was a distinctly competitive atmosphere at the Show. Concorde was being put through its paces by one of the French test pilots, Jean Franchi, and there is no doubt that he put on a very graceful performance. He did nothing which was not entirely within the capabilities of Concorde, but he was throwing it around quite a bit, landing and then taking off again within the length of the runway and generally showing off its capabilities. Whether the Russians thought they had to match this exhibition there is no way of telling, but if they did it was a silly thing to do with an expensive piece of equipment at an air show.

Responsibility for investigating the accident lay with the French, because it happened on their territory, but you need the full technical back-up of the manufacturer if you are going to produce a conclusive report. No flight recorder was found, or was ever reported as having been found, so the accident is likely to remain a mystery. It is important to note, however, that the crash had nothing to do with supersonic airplanes as such, and was in no way indicative of some inherent fault in the type.

That apart, the Russians have never proved to be real competitors outside the Eastern bloc of countries in the civil airplane business. In the limited number of cases where they have sold their airplanes abroad, such as Egypt, India and Ghana, the tale has not been one of success. With them developing their own model we had the best of both worlds. We did not have them as major competitors in the West, but at the same time we did not feel all the loneliness of the long distance runner, because there was at least somebody else padding along behind. We had always given a wary welcome to the American SST programme because it would have helped to solve the lingering problem of landing rights and mutually acceptable airline schedules. The story that the TU144 is a direct copy of Concorde, based on industrial espionage work, must be laid to rest. I would like to repeat and re-emphasize this point. The similarities are relatively superficial. If you look at the shape of the wing and especially at the engines and intakes, there are major differences between them. In a way, aircraft designers are rather like dress designers; couturiers in Paris, Rome and London, without any collusion at all, do come out with similar styles in a particular year. At any given juncture, if you were to ask aircraft designers in California, England and Russia to produce an airplane which fulfils a particular requirement, the state of the art in all three places tends to be similar enough for the designs to be likely to look much the same.

After all, the McDonnell Douglas DC9 is almost indistinguishable to the layman from the BAC One-Eleven, although there are a number of major differences between them. The DC8 and the Boeing 707 are almost indistinguishable, and so, apart from the rear centre engine installation, are the DC10 and the Lockheed 1011. This is not because industrial spying has been going on but because the state of aeronautical knowledge inevitably leads people to a common optimum solution.

The Russians have never made any serious attempt to compete with Concorde abroad. There was one period when they put the odd advertisement in a professional magazine like *Aviation Week*, but they have made no serious attempt to sell their aircraft in the West. The Russian aircraft industry makes its airplanes primarily for Aeroflot, and if they happen to sell some elsewhere that is a bonus. Aeroflot needs a very large number of airplanes. They have a huge area to cover.

11 Open Government

With the election of the new Labour government in February 1974, those of us connected intimately with the Concorde project endured the most nerve-wracking, and in many ways most bizarre, crisis of all. We had ample premonition of danger. The country had just survived the miners' confrontation with the Conservative government and the effects of the three-day week. The economic situation had deteriorated sharply, and it became apparent to us within a matter of a few days of the new government taking office that Denis Healey, the Chancellor of the Exechequer, as part of the action that he had to take to deal with the economic crisis, was going to try to carve up the Concorde project. It was all starting to look like an ominous echo of the 'Brown Paper' crisis of almost exactly ten years earlier.

Another essential element in the sequence of events for the next three months was that Harold Wilson had divided the former Department of Trade and Industry into two separate Departments, with Anthony Wedgwood Benn at the Department of Industry, which included responsibility for Concorde, and Peter Shore becoming Secretary of State for Trade, which included responsibility for British Airways. This meant that instead of having one minister responsible for both, to which we had recently become accustomed, we were faced by two parallel lines of communication.

Although that aspect of it did not please us, we felt strengthened by having Tony Benn on our side. I had had a fair amount to do with him in the latter part of the Sixties when he was Minister for Technology. He has always been an advocate of Concorde, and we had had several useful discussions. He represents a Bristol constituency, although not the one in which our works at Filton lie, and therefore has always been thoroughly familiar with the local social and employment scene, and he had been of great assistance to us when we were trying to launch the BAC Three-Eleven.

I had always found Tony Benn one of the most intelligent and reasonable ministers with whom I have ever had to deal. Many of my friends throw their arms in the air in amazement when I say this, but I can only state it as a fact. I found him quick to take a point, quick to understand what you were saying, reasonable and searching in his questions, and extremely fair in his judgements. I believe that Sir George Edwards, whom I would have thought most unlikely to share Tony Benn's political opinions, would give the same appraisal. As a minister, Tony Benn is a joy to talk to because he is so much brighter than most of the ministers with whom one has to deal.

I state this as a preliminary because we were certainly not prepared for what happened on 18 March 1974 and were badly shaken by the event and its aftermath. At half past three that afternoon, Tony Benn made a statement in the House of Commons about Concorde. We had had no prior notice of it. There had been no previous discussions with Sir George Edwards, and the copy of the statement was sent round by hand more or less simultaneously with his standing up in the House. The form of the document was itself curious. It had the government crest at the top left-hand corner, and the superscription 'Secret' had been crossed out by hand, with the words 'Unclassified after 3.30 pm 18 March 1974' written in by hand underneath it.

The first page was an innocuous enough summary of the current production situation. It gave the current figure of the joint development costs as £1070m, of which a further £130m remained to be spent in the UK. The first paragraph on the second page was ominous: 'British Airways estimate that the operation of Concorde could substantially worsen their financial results, possibly by many millions of pounds a year.' There were unfavourable references to the noise level compared with new subsonic aircraft, an estimate of a selling price for the plane of up to $47·5m, which it was stated would not be sufficient to cover the cost of production and the necessary market support measures, and a further calculation that the United Kingdom's share of production losses, depending on the number of aircraft sold, could range between £120m and £300m. 'In view of the size of sums of public expenditure involved and the importance of the decisions that must now be made, I thought it right to place all these facts before the House and the country before any decisions are reached,' the statement concluded.

This came as an absolute bombshell. Our telephone lines were choked with Press inquiries and the BAC top brass got together to decide what to do. It looked like the end of the line, as the general tone of the document was doom-laden. I brooded on the reference to British Airways and said to Sir George: 'I wonder if British Airways, having seen that, are thinking about issuing a statement, because the one thing we don't want is for us to get into a public slanging match.' So I called up Henry Marking, who at that time was the Managing Director of British Airways and is now Deputy Chairman. 'Henry,' I said, 'what about this thing from the Department of Industry? We do not propose to burst into print immediately and think it would be better if we sit and stew over it for a bit and probably have a word with you.'

'Oh,' he said, 'we've already issued a statement,' which indeed they had. It referred back to the agreement in 1972 by which they had ordered five Concordes with a guarantee that the government would subsidize any loss in operation. It went on to say that their 'estimates range from an improvement on profitability of £6m a year to a worsening of £26m a year'. The statement continued: 'Unfortunately the latest estimate of the effect of operating five Concordes is, at best, towards the most adverse of that range of forecasts. Greatly increased fuel and other costs affecting all British Airways operations and uncertainties of Concorde route permits, contribute to the situation.' That really put the fat in the fire. We took the decision that we were not going to leap into print, and Sir George only issued two very guarded statements the following day and later in the month, declaring his belief in Concorde and urging everyone to spare no effort to get it into passenger service as quickly as possible.

Nevertheless, we feared the worst, and wondered what professional and public opinion in other countries were making of this altercation. The British really can behave in the most extraordinary fashion. There we were, having spent the immense sums of money involved in the technological marvel of Concorde, busily engaging in the circulation of public statements designed to make the product unsaleable.

By 27 March, BAC had put together a rather laborious statement in refutation of the main charges and, in conformity with what we assumed were Tony Benn's conceptions of 'open government', issued it to the Press. We disputed British Airways' calculations, asserting that in our judgement Concorde could be operated profitably with a sixty per cent load factor, and we questioned a whole range of production and operational assumptions. It was not a very dynamic document, so in order to build up our defences we tried to find out what had been going on.

The roots probably lay in Anthony Wedgwood Benn's intervention in the House of Commons debate in December 1972. He complained that there had been insufficient parliamentary scrutiny of the original Concorde agreement with the French, and declared that there had been a cloak of secrecy over the details ever since. He admitted that he had played his own part: 'I was dealing with the cloak of secrecy which surrounded the aircraft from the outset. I share responsibility for it, though in self-defence I might say that when my Rt. Hon. friends and I were dealing with it there was a competitor, which there is not now.' Since then he has made a fetish of what he calls 'open government' – making all the facts about Concorde available, with ministers not just answering questions on the basis of carefully prepared briefs by their civil servants, but making all the information available to the public.

The Treasury has never liked Concorde. It is their way not to like expensive prestige projects, and there was a substantial body of opinion in the Labour Party which would always prefer to see sums of this magnitude spent on social projects rather than on an aircraft to carry first-class passengers.

Tony Benn, who has always been a supporter of Concorde, doubtless sensed this mood, and on taking office requested his senior civil servants to prepare an up-to-date brief on the latest statistics. It was essentially a summary of this report, still on Civil Service paper, which had been distributed on 18 March. At first sight, it seemed a surprising action for a minister to take who was known to be in favour of retaining Concorde, and it was strange that a statement from the Department of Industry, which was the sponsoring department for the project, should be as damning as it was.

There was another factor that disturbed us. For many years the two senior civil servants responsible in the Department for the project had built up a body of experience on Concorde as great as that of many of the people associated with it in BAC. They knew the background well and had lived through all its vicissitudes. Shortly before the Labour government came into office, both of them, such is the way of the Civil Service, were transferred to other duties. At this critical stage in our affairs, we found ourselves dealing with new people at the Ministry who had very little in the way of relevant past experience to guide them and every reason to be looking over their shoulders at the mandarins in the Treasury.

The ministerial document explained as plainly as you can in parliamentary language that although there would be a high price to pay for cancellation, there would be a higher price to pay if the project were not cancelled – that in real terms it would cost the Treasury more to go on, in spite of the fact that we had very nearly finished development in money terms. It also embodied the assertion of British Airways that they were going to be a good deal less profitable if they operated Concorde than they would be if they did not.

This did not mean that Concorde could not be operated profitably in airline service. The assertion was rather that for an airline with a fleet of Boeing 747s and other subsonic jets, the effect of removing the high-yield first-class business from the 747s and attracting it to Concorde would be to make the 747s unprofitable. Such a shift would mean that, overall, the airline would make less money than if it did not have Concorde in its fleet at all. This argument was not understood by most people, and it is the critical point, as Concorde by itself can be operated profitably. It was the effect on previously acquired aircraft which in the estimates of BOAC altered the picture. The BAC argument was of course that when the Boeing 747s were ordered the airline was well aware that Concorde was coming along at a later date, and that this should have been kept in mind at all times.

We immediately arranged for a high-powered BAC delegation to have a session at the Department of Industry with some embarrassed civil servants. In fairness to them, we believed that they thought they had simply been producing a brief for their Minister, and had not expected it to become the first example of his open government policy. Indeed one senior civil servant in the Department said to me at a later date that the effect of the publication of the statement had been beneficial in one sense, for it had cut down the amount of paper work flying round the Department and made people think more carefully about what they actually put down on paper. It was agreed that we should have full opportunity for examining and querying the figures and reporting back at an early date. This gave us a certain amount of leeway, although for the next couple of months the general atmosphere was one of gloom and despondency.

Curiously, this alarm had very little effect on our work people at Bristol. They had been working on Concorde for so long and had seen so many crises come up that their reaction was: 'Concorde has been through all this before and it will survive again.' As it turned out they were absolutely right. Those of us who were a little closer to the centre were, however, more worried about this particular crisis than almost any other we had been through.

So why did Tony Benn take this initiative? I have a considerable respect for his ability as a political operator, and I have no doubt that he saw in this precipitate publication of his officials' report a way of staving off what might well have been the alternative, an early Treasury-backed decision to cancel the project. I have since talked to Tony Benn in calmer times about the episode, and although he is very careful not to reveal any Cabinet secrets I doubt whether my interpretation goes far astray. He dismisses the curious form of the original statement but does insist on the point that, by putting the problem to the public before any decisions had been made, when the Cabinet finally did make up its mind about Concorde some weeks later it had had the benefit of formidable representations by those most concerned, so that it was the best-informed debate on Concorde that there had ever been.

Tony Benn says:

> This was in line, I won't conceal it from you, with my view that decisions of this magnitude ought not to be made and announced but should be discussed and then be made and announced. To that extent it may have played some part in stimulating public support for Concorde, but it was in part a response to the fact that successive Select Committees have criticized ministers, including myself, for having failed to publish adequate information before decisions were reached on Concorde.

He is properly cagey about the Treasury attitude at the time, but he does make the point:

> To put it crudely, the Treasury is always in favour of saving money. The Foreign Office is always in favour of good relations with foreign countries. The Department of Industry is always in favour of British industry. The Department

> of Employment never wants to see people sacked and the Department of Health and Social Security always wants the money that's available to be spent on pensions and hospitals. You can imagine how any Cabinet of any party will react to almost any issue, because people do have a departmental interest and, of course, the Treasury, from the very early days, has always had great grave doubts about expenditure of this magnitude on single projects.

He dismisses our own fears that the report he received was in any way influenced by Treasury thinking or that the British Airways attitude, which would have come to Cabinet from Peter Shore, was in any way encouraged by those who opposed Concorde. No conspiracy, in other words. He also stresses that his action had the full support of the Cabinet:

> It was a very big issue. The normal way in which governments proceed, of course, is to get their information from the industry or the market or whatever it is, consider what to do about it and then make a statement in which the decision is announced. What we did in this case was to make the information available. If you look at the statement very carefully, the only thing that was surprising about it was that it revealed that the government had received a very gloomy report about Concorde. It would have been a much bigger bombshell if we had followed normal practice by announcing that the project was going to be cancelled. I think the Cabinet was quite right to take the view it did, that you must give people an opportunity, if there has been half a generation of work done on a project, to express their own view before the government reaches its final conclusion. It did play some part of course in saving Concorde, there's no question about that.

What then happened was that British Airways were asked to make a full report on the situation as they saw it to the Department of Trade. This they did at some speed. In fact, the letter of 9 April by their Chairman, David Nicolson, to the Minister, was a comprehensive but not very favourable document, which stuck to their estimate of about £25m per annum as the cost to British Airways both of operating Concorde and losing first-class subsonic traffic. BAC got a copy of the report and the letter, which included this paragraph:

> Nothing has happened to eliminate any of the uncertainties on which our original policy was based. On the contrary, the uncertainties are now greater: indeed, for instance, we do not yet have guaranteed rights for operating one single route. Moreover, the fundamental changes in the economy of world airline operations brought about by the energy crisis – above all the huge increases in aviation fuel prices – have created immediate and perhaps prolonged difficulties in maintaining profitable subsonic operations.

It was their and our clear understanding that this report was not to be made public, and a very strong request was made to the Secretary of State to that effect on the grounds that it was best considered confidentially, because if people were going to make reports to a minister which contained the whole situation as they saw it, they were more likely to be frank and honest if they did not think it was likely to end up in the newspapers. We then had a high-powered conference with Peter Shore in the chair on 17 May, at which we tried to reconcile the widely differing interpretations of BAC and British Airways. It did not solve much, and ended with the Secretary of State saying that it would be useful if BA and BAC could hold further discussions to try to resolve some of the differences between them.

To our considerable annoyance, we were informed by the Department of Trade on 30 May that they proposed releasing the whole documentation to the Press the following day. Whether they were trying to knock our heads together I do not know, but it seemed to us to be a notable lack of courtesy. We also felt that it could be extremely damaging, as we had not yet resolved our differences with British Airways.

Now no one is going to believe that what follows was a coincidence, but it so happens that, on the very day we got notice from the Department of Trade that they were going to release the documentation, I had arranged to have lunch with Freddie Laker, whom I had not seen for about six months. I wanted to discuss with him an enthusiastic report he had prepared and sent to us for analysis, in which he proposed a scheme for the profitable operation of Concorde. However, it was not until I got back to the office after lunch that I was told about the Department of Trade's release. I suddenly thought to myself: 'What about Freddie Laker's report that we've been sitting on?' Freddie had gone off to his dentist, and I did not have the faintest idea how to get in touch with him. However, I finally tracked him down at one of his London offices and Freddie responded immediately: 'If everyone is going to release their report, why don't you release mine? Why sit back and allow a damaging report to be put out to the Press? Why don't we make sure that equal coverage is given to this very thorough study of mine?'

So Freddie and I sat in his office until about 9.30 at night setting up a Press conference for the next day. Freddie is always extraordinarily good news value. Once he decides to go into action, people have to listen to him because, although he is regarded as a bit of a maverick, he is quite the opposite. He is far and away the most expert airline operator in the country, and consistently, year in, year out, he has made money.

The result was that the British Airways report hardly made the front pages. Freddie Laker was splendid. 'Certainly I can operate profitably,' he told the Press. 'I've got a long record in the aviation business, longer than almost anyone else. I have never lost money in my life. This is not just a decision I've taken off the top of my head. I spoke to BAC back in 1973 and got them to do a lot of evaluation work for me. I've done my own studies, which were made available to BAC some months ago, and they showed that I can certainly make money with Concorde.'

In answer to the question: 'Would you be prepared to put

your money where your mouth is?' Freddie Laker answered: 'If the government is prepared to make the cash available to me to buy the airplanes on precisely and exactly the same terms as it is prepared to make money available to British Airways, then yes, I am ready, willing and able to go.' The British Airways statement was completely trumped, because Freddie Laker is immeasurably better news value than almost anyone else in the business. So he got the headlines and, without doubt, did a very great deal indeed to restore the viability of the Concorde project in the eyes of airline people and much of the public at large.

Tony Benn then continued to play his part. 'I pumped all the comments that came in back into discussion in government,' he has told me since, 'so that the Cabinet had the opportunity of seeing what other people had said about figures that otherwise they would have received simply through the normal private channels of direct contact between the airline and the Department of Trade.'

So what conclusions have we to draw from this whole bizarre episode? One thing we have learned over the long gestation period of the Concorde is that if a government starts thinking about cutting the project's throat, the only way it will get away with it is if it is done quickly. The more the weeks and months drag on, the more difficult it becomes. It is not entirely clear to me why this is so but experience tells us that that is the way things tend to work. While we did not set out deliberately to delay the decision, we put forward all the arguments and mounted all the propaganda for retaining it. The effect of Tony Benn's intervention was to reinforce this. One of the senior civil servants in the Cabinet Office said to me some weeks later that although Benn's methods were extraordinarily unorthodox, they were also extremely effective.

You will not find a record of any meeting where Tony Benn staked his political life on Concorde going ahead, although in some ways he was doing just that. The statement that he released, based on his civil servants' report, seemed to do more damage than anything else to Concorde. Yet, from that starting-point of so-called 'open government', he got us down to the job of saving the project.

He is by inclination a bit of a technocrat, as well as an efficiency expert. Technical problems appeal to him. He understood about the airplane and, what is more important, its economics, which is really what the arguments revolve around. Many people would say that we are very fortunate that he is a Member for a Bristol constituency, and that is what it is all about. I am not convinced of this. I think many of his ideas about shop stewards participating in the management of plants and the grass-roots trade union business do come from his Bristol associations, and specifically from his Rolls-Royce and BAC links there. Some lecturers in Bristol University, in conjunction with shop stewards in our plants, produced a document about this time on how the industry should be nationalized. Tony Benn encouraged its preparation, and it has his picture on the front. Don't forget that he was not in Parliament at the time Concorde began, because he inherited his father's title in 1960 and did not get back in again until 1963, after he had disclaimed it. His personal political strategies and his support for Concorde do not necessarily derive from the same root.

I can only record the opinion that if we had had a Labour minister less effective than Tony Benn, Concorde would definitely have been cancelled in 1974. I cannot land him with this judgement, but I have no doubt that he set out to save this project and won a victory over Denis Healey and the Treasury in the process. He saved the entire development and production programme up to the limit of sixteen production planes that we are now building.

He grins disarmingly when you challenge him with this and says:

> I don't really believe in the role of individuals in quite that sense. I will say this. Having made it clear that there was a big decision to be taken, the Cabinet was able to take it rather more easily because it had invited comment to get

a full assessment of what the consequences would be of alternative policies. I think I did help the Cabinet by getting agreement on this, to see the merit of having a discussion about the project and not to make the mistake of reaching a view without hearing what everybody's views on it really were.

I think it would have been quite wrong to have taken a different view. We shall see within a matter of weeks or months of the aircraft entering service how right we were, but having spent so much time and effort on it, it was absolutely right to continue production. I never really varied from that view.

And what did our friends in Toulouse make of all this? Well, we got a number of anxious telephone calls, and if Henri Ziegler had still been in charge I do not doubt he would have had a fit. But the general reaction was very different to what it would have been ten years before. They had learned from experience that this was one of the extraordinary things that happened, particularly when the British had a change of government. They were involved at the time in a number of demonstration flights, which were carried out on the score that the more the airplane was taken to different parts of the world, the more it would emphasize that the project was going ahead. They took their pre-production Concorde 02 to the West Coast of the United States with great success and appeared, at least on the surface, to remain remarkably calm.

The plane itself was facing a new set of problems. Technologically it was perfect. It was doing everything planned for it a dozen years earlier. The production planes were coming out of the two factories in an orderly fashion. But we still had to contend with the world environmentalist lobby, and we still had to get clearance from its destinations in regular airline flight.

12 Intelligent Atoms

Hardly any major industrial or technological project in the modern age has had to suffer the barrage of criticism that has assailed Concorde throughout its life. There have been endless articles in the Press questioning every single aspect of its development and operation. Government inquiries and parliamentary debates have taken us up by the roots and replanted us, which is fair enough, because it is public money that we have been spending. Many books have been written to prove that Concorde is a technological, ecological and financial disaster. But the intensity of attack has always baffled and surprised us.

One of the earliest campaigners was a gentleman named Professor Bo Lundberg, former Director General of the Aeronautical Research Institute in Sweden. He provided much of the material for a book called *Concorde – The Case Against Supersonic Transport*, by Richard Wiggs, Chairman of the Anti-Concorde League, which was published in 1971, and also for another called *The Concorde Fiasco*, by Andrew Wilson, the aviation expert of *The Observer*, which was brought out as a Penguin Special in 1973. Lundberg originated many of the ecological and environmental charges which have been laid at the door of Concorde, and the other two took them up and embellished them with an array of arguments about Concorde's technical specifications and economics in airline service.

Wiggs seems to be something of a self-appointed campaigner, although he has attracted a certain number of subscribers to his Anti-Concorde Project. He gave a Press conference when he launched his book and one of the aviation correspondents whom I know well asked him: 'What are you going to do, Richard, if you win and the Concorde project is cancelled?' To which Richard Wiggs is alleged to have replied: 'There will be other barricades to man.' So I suppose he is one of those people appointed to set the world to rights and keep us on our toes.

There is another name I should interject before I deal with Andrew Wilson, and that is Mary Goldring, who during the first few years of the Concorde enterprise wrote highly critical articles about it in *The Economist*. She is an extremely intelligent and forceful character, with whom I was obliged to differ on a number of occasions. We had one tremendous knock-down, drag-out argument on BBC's Panorama, when we were very tough with each other – although we got on personally perfectly well. When she left *The Economist* at the time that my namesake, Andrew Knight, took over as Editor, we had lunch together to reminisce about our past differences. Anti-Concorde observers would say that she had mellowed with time; pro-Concorde people might claim that she had changed her views over the years and adopted more sensible attitudes. It entirely depends which end of the telescope is being used.

I have never really been able to fathom Andrew Wilson's hostile attitude to Concorde. It seems to me that his articles on it could seldom be called dispassionate. I understand that he was entirely supported by his Editor on *The Observer*, David Astor, and I have the strong and clear impression that it is a campaign which is being waged, a total dedication to stopping Concorde. It is hard to single out his objections, because he operates over the whole spectrum. If they are environmental, it is difficult to discover which environmental arguments he considers substantial, because he covers the lot. Does he consider that the airplane is uneconomic, that the country cannot afford to spend this sort of money? Nobody knows, because he attacks right across the whole front. At no time has he said: 'My real reason for objecting to the airplane in essence is this...'

I would entirely understand if Wilson were a total believer in the environmental evidence that has been produced, or if he consistently attacked the noise problem, whether it was due to Concorde or not, but he has not done that either. If he was a consistent supporter of the economics of air travel and sincerely believed that Concorde was an uneconomic vehicle, it would be possible to refute his case. But none of his writing shows a passionate belief in either of these aspects: he appears to be just generally anti-Concorde, which is very strange. He has reported on occasion from the hearings held from time to time in Washington and New York about the operation of Concorde and supersonic transports. Yet you would never know that a great deal had been said in support of Concorde at these hearings.

At one point in 1972, I became so incensed by one of Wilson's articles in *The Observer* under the title 'The Fastest Flop on Earth', that I wrote a letter to the Editor to protest. I found it necessary to say: 'We resent and deplore the antics of *The Observer*. It may be great fun for *The Observer* to use its privileged position to denigrate Concorde by publishing in the guise of facts misinformation which is without foundation, and by drawing conclusions from half-truths which are wholly unwarranted. We who work on the project do not find it funny.'

On that occasion they gave me the right of reply, and I challenged Wilson's figures. But it is difficult to win an altercation with a newspaper, for not only does the Editor usually decide to have the last word but he also decides exactly where articles are positioned in the newspaper and hence what weight they carry. On the day my article appeared, Concorde was subject to unfavourable editorials in both the main and the business sections of *The Observer*, and furthermore Wilson was given the opportunity to have the last word in the next Sunday's edition. At that point I gave up.

Nevertheless since these criticisms have provided a constant accompaniment to Concorde over the years, I should perhaps refer to some of them here. They fall into two main categories.

Firstly, the science fiction element of damage to the ozone layer, the threat of cosmic rays and the destruction of the stratosphere by jet engine exhaust fumes – ideas which mainly derive from Bo Lundberg. Secondly, a tangible category which is concerned with the airplane's noise level in operation. This is in its turn divided into two elements: the noise on take-off and landing, and the effect of the sonic boom which accompanies the aircraft in supersonic flight.

In the first category it has to be said that nothing whatsoever has been substantiated. We took these arguments very seriously and made a full scientific exploration of the problems with both European and American research teams. Although it would be improper for someone in my position to say that the suggestions are nonsense, the fact remains that the reports are entirely negative. I think that this kind of argument is similar to that made when railways were invented in the middle of the last century. Prince Albert, the Prince Consort, who was considered to be a scientific gentleman, subscribed to the theory of certain London doctors that, although trains could travel at sixty miles an hour, they should not be allowed to do so because no human being would be able to breathe at that speed. Scientific advance does tend to breed that type of argument.

It has also been alleged that supersonic operations in the stratosphere will break down the ozone layer which protects the earth against ultra-violet light – and so cause serious disturbance to the natural balance and structure of the atmosphere. This will, it is claimed, produce radical changes in the earth's climate and even cause skin cancer. It has been further alleged that the ozone layer will be destroyed by the oxides of nitrogen from the jet engine exhaust of high-flying supersonic aircraft.

Credence to these claims has been elicited by many pessimistic forecasts, especially in the United States, by those whose objectives were by no means always of pure scientific enquiry.

When the subject was raised in the United States, Henri Ziegler, my Aérospatiale colleague, would turn to his questioner and flatter him by saying: 'Well, you people have tremendously advanced technology. You have even invented the intelligent atom.' When they asked him what he meant, he would say: 'You now have intelligent ozone atoms in your atmosphere. They sleep quietly protecting the world and when a supersonic military aircraft passes they wake up and look around and say: "Oh, it's a US bomber." They go to sleep again and keep protecting the world. Then there is another bang which wakes them up again: "Oh, it's Concorde" and they raise all hell.' This was a good answer to his American audiences because supersonic military aircraft have been flying all over the world for a couple of decades now, and fly in some numbers every day in the United States.

Sir Archibald Russell was even more terse on this subject. He holds to the opinion that ascending gases from belching cows had more effect on the ozone layer than supersonic aircraft and Sir Stanley Hooker considers that aerosols had more effect on the stratosphere than Concorde. I do not wish these refutations to sound too flippant, but these people are experts and base their conclusions on profound scientific study.

Apart from the fact that there has long been a great volume of aircraft operation in the stratosphere, by both commercial subsonics and military supersonics, which has produced no discernible adverse effects on the climate, these alien forecasts have now been substantially refuted by some of the most eminent scientists and scientific evidence in several countries. Their conclusion is that it will be impossible to detect the variation of the ozone amount so caused within the naturally occurring variation – which can be of the order of as much as thirty per cent over a period of as short as a few days. Moreover, despite periodic injections of oxides of nitrogen by nuclear weapon tests, and the increasing volume of stratospheric operation, the amount of ozone in the stratosphere as measured at several stations around the world has, in fact, been steadily increasing. In some cases by more than 0·5 per cent per year.

Our overall conviction is therefore (and this is now shared by many responsible bodies) that, analysed scientifically and mathematically rather than emotionally, there is little or no evidence to support any of these allegations, and that monitoring will, in any event, provide an absolute safeguard. I'm afraid Professor Lundberg and other like minds will have to produce very much better evidence than they have so far if this doomsday talk is to have any relevance.

The noise problem is the most recent on which the environmentalists have been building up a determined case. There are two aspects involved here – which must be considered separately – the sonic boom and airport noise.

The sonic boom phenomenon is the principal problem associated with supersonic transport. However, it is the lesser of the two noise aspects because everyone concerned has long agreed that sustained flight at supersonic speeds will only be permitted over the sea and over land areas of sparse population such as deserts – which together form a considerable part of the earth's surface. Hence very few people indeed would need to be affected. In this context, it is significant that around eighty per cent of today's intercontinental seat-miles are, in fact, flown over oceans or land areas of this kind.

It would be a great advantage from the sales point of view to be able to fly the Concorde from Frankfurt to New York nonstop, but the first part of the operation is overland – Germany, Belgium and France. The cost of operation is almost exactly the same to fly subsonic as supersonic, because the aircraft burns the same amount of fuel per hour. But it is only going half as far subsonically as it does at its full supersonic capability. Hence the cost per mile over that phase is twice as high. So it is a matter of economics, and obviously makes Concorde somewhat less attractive, if it has to fly very significant stretches subsonically over land. Whether or not 'supersonic corridors' are socially acceptable over particular countries will be a decision for their governments to take in the light of public opinion and mutual negotiation. However, widespread practical evidence has shown the boom to be far less intrusive than many of the more emotive commentators would have us believe.

As far as airport noise is concerned, we originally set ourselves the target that on entry into service Concorde's would be of the same order as the then current subsonic jets, such as the Boeing 707 and the McDonnell Douglas DC8. (Even now, large numbers of these first-generation jets continue in front-line service and will still represent well over two-thirds of all departures at international airports throughout the world for many years after Concorde's introduction.) While the Concorde prototypes without the full production standard of silencing devices were undoubtedly above this level, we have since demonstrated that we have achieved our objective.

However, there are now new international noise requirements, known as the International Civil Aviation Organization (ICAO) Annexe 16, and the American Federal Aviation Regulation (FAR) Part 36, which define the noise levels that will be normally acceptable at airports in the future. Concorde noise is broadly the same as that of the first-generation subsonics. Thus ninety per cent of all the aircraft operating with airlines today are above these new rules as well.

A supersonic airliner must have a straight jet engine, which is bound to be more noisy than the newest subsonic types. These are quietened by having a bypass system which slows down the jet of air and gas, and with the engine taking more air than the compressor and the turbine themselves need. The air that goes around the engine core is mixed with that which passes through the engine itself, so that the efflux mixture is substantially slower than that of the straight jet. The fan at the front of the subsonic engine is getting progressively larger, so that it is blowing more air around the engine in addition to that which goes through it. The velocity of the jet efflux is thus very much lower, and the aircraft very much quieter. There are still great potential opportunities in subsonic planes to reduce engine noise even further. The second-generation jumbos are quieter because they have enormous fans fitted to the front of their engines, which really makes them into a kind of shrouded multi-blade propeller with variable pitch. In the supersonic regime the momentum drag of the engine depends on its frontal area and any larger diameter for the powerplant installation in Concorde could not be afforded or accommodated in the present design.

Sir Stanley Hooker and his people from Rolls-Royce got together with the French and other experts in the late 1960s to form a noise panel to study the problem. This was a most complex task, because it necessitated charting the whole physical mechanism of jet engine noise generation. What is the noise frequency spectrum and exactly where in the engine machinery and combustion system are the sources of noise? It was thought that if these problems could be identified some practical methods of reducing the noise might emerge. The only really worthwhile means of attenuation found was to squeeze the jet out with the thrust reversers so that it spreads horizontally. While not as satisfactory a solution as the bypass principle, at least it was achievable.

There is a solution for the future, but it would make Concorde twice its present size and weight. In a bypass engine the thrust per pound of air ingested is more or less halved, and with twice the airflow the intake has to be twice as large and twice as long as that of the present Olympus. Only a second-generation Concorde would be able to accommodate this.

There is a redeeming element. Concorde climbs much faster and at a steeper angle than subsonic planes and removes its noise from ground level in a shorter distance and more quickly than they do. Nor should it be forgotten that at an airport like Heathrow there will probably be not more than a handful of Concorde flights a day in the initial stages, compared with the hundreds of take-offs and landings of all other jet aircraft. Moreover, the agitation seems to have a national rather than an international flavour. American objections can never be entirely divorced from their failure to produce an SST of their own.

The published complaints in Britain have come almost entirely from the anti-Concorde organizations – which have made it their business to take peripheral measurements – and most of these have come from those easily identifiable with the Noise Abatement Society, who are trying to make a case against all aircraft noise. In practice, I am not aware of any general, popularly based, storm of protest from around Heathrow.

In France, which has shared the production of Concorde, there has been to all intents and purposes no measurable protest at all. On the demonstration and proving tours that Henri Ziegler and I went on around the world, I do not recall a single instance where there was any complaint about the noise that Concorde made coming in or taking off.

When the Olympus engine was originally conceived more than twenty years ago aircraft noise was not a matter of public concern and there were no airport noise regulations in existence. With the subsequent huge development of air traffic and proliferation of the jet generation, I would be the last person to say that aircraft noise does not matter and that people will just have to put up with it – and the groundswell of public opinion that has generated the new noise regulations is testimony to this. But we have got ourselves into a very curious situation which was reflected in the long-drawn-out controversy during the Sixties and Seventies about where to put London's third airport before the Maplin plan was finally rejected. It is worth bearing in mind that when the John F. Kennedy Airport in New York – which is being as searching about Concorde noise as anywhere and where local politicians have to pay close attention to what the public have to say – was originally built under the name of Idlewild, there was no one there at all. The airport land area was reclaimed from the sea, and all the houses which now contain residents who complain about the noise moved in after the airport was built and because it was there. The same is to some extent true of London's Heathrow, which was out in the countryside west of London when it first started to be used.

I think it perfectly right and proper that, if technology can achieve it within reasonable economics, aeronautical manufacturers should be obliged to build quieter aircraft. I also think that automobile builders should be required to produce quieter motor cars, trucks and buses – and less smelly ones. While further small improvements may be achievable, it is no use anyone suggesting that we have to quieten Concorde substantially from the airport noise point of view because with the initial production model this just cannot be done without very considerable further outlay. It is my belief that after all the verbal emotion has subsided the human ear is not really going to detect sufficient difference between Concorde and the continuing masses of Boeing 707s and DC8s to make a real issue of the scaremongering allegations that have been deliberately built up by the antagonists. Unfortunately, however, because of Concorde's distinctive 'visual signature' people will readily be able to identify it apart from all the subsonic jets and this, unhappily in my view, will by itself lead to complaints being made when they would otherwise not have been if Concorde had looked like all the rest.

Another attitude which has played its part in the reaction to Concorde is the rather unpleasant sense of envy which has grown up in our political life since the war. It produces the attitude that Concorde is somehow an aircraft for rich businessmen and that all the rest of us have to do is to put up with the boom. Henri Ziegler has as good an answer to this as anyone. Although the plane has enjoyed general approval in France, during the whole period of its manufacture, it did come under a certain amount of criticism in the spring of 1971, principally from a magazine editor who was also a member of the French Parliament, Jean-Jacques Servan-Schreiber. At about that time, a former French Prime Minister, Antoine Pinet, was sufficiently impressed by some of this material to refer to Concorde as 'an aircraft for millionaires'. Henri Ziegler reacted quickly. The same day he appeared on radio and television to say: 'We are now in competition with the Tupolev 144 and they plan to make their first flights between Moscow and Calcutta, that is to say between the capital of all the Socialist states and one of the poorest places in the world. So I would rather say it is an aircraft for proletarians.' He never heard anything more about Concorde being an aircraft for millionaires. He also, as we did in BAC, enjoyed the whole-hearted backing of the unions in the project, and they came out strongly in support, even the CGT, which is the Communist union. Pinet was abashed by what he had said and apologized.

Is the agitation in Britain about Concorde noise and all the other popular protest propaganda an expression of an unfortunate development in the national psyche? In the thirty years since World War Two, Britain has slipped downhill from being a large nation with an Empire and expansive world ideas into a poor nation looking inwards, unwilling to embark on major national projects, complaining querulously about minor aspects of those that do reach fruition. France has greater national pride in prestige projects embarked upon with the full approval of the French people; there is little argument against them. I ascribe the difference in the two countries not only to national attitudes but also to the different economic conditions that have developed since the war, with France prosperous and Britain in decline.

Without trying to minimize the shift in public opinion, one of the regrettable results of not getting Concorde into airline service at the beginning of the Seventies is that it would have been accepted as being no more noisy than the subsonic jets then in operation, as it is only with the newer versions of these jets that it has been possible to bring the noise level down marginally. Nevertheless, we are going to have very severe problems if the decision finally taken in the United States is to keep Concorde out of New York. It is exasperating that many of the arguments put up at the US environmental hearings are highly emotional assertions not really based on facts. The environmentalists are putting up some pretty incoherent arguments, which we cannot accept, and they appear unwilling to accept the facts we place before them. Decisions may be taken for the wrong reasons, and if New York is out, then obviously the potential for Concorde becomes substantially reduced.

Even if it can fly into Washington Dulles Airport, this is not the same from the international air traffic point of view as flying into John F. Kennedy Airport, New York.

In summary, the plain truth about Concorde noise is that it is no worse than the noisier current American jets which are in use in their thousands all over the world and will continue to be in use for many years to come. Concorde's 'sin' is that it is no quieter either. But as a bonus, Concorde will get to its destinations in half the present times, and that great step forward has been achieved without worsening the environment at airports or anywhere else.

I have heard Anthony Wedgwood Benn make a point of substantial consequence here, which I think it fair to record because it seems to have caught his politician's eye rather more quickly than it did some of the rest of us:

> The technical achievement of Concorde is absolutely outstanding. You've produced everything you said you'd produce – a little late and at greater cost – but from the engineering point of view and the degree of safety and testing, absolutely brilliant. But what nobody really noticed because no one was responsible for looking at it, was that at the time the aircraft was born, speed was everything and by the time it has been produced, the environment is everything.
>
> This subtlety of change in public mood was never studied with the same degree of skill and care and attention as the engineering problems, and this factor has created very great difficulties, to the point of changing the prospects for the project. People now want peace and quiet. I'm not saying they do not also want speed, but the environmental arguments have played a considerable part, not least in the cancellation of the options.
>
> I think if I have criticism of BAC, which I would not make of the French, it was that they have never really realized that when you are building an aircraft with taxpayers' money, with its acceptability depending on public opinion, not just the view of the airlines who will fly it, you simply have to

operate on a completely different level. They were, I think, tending to say too much: 'We're producing a piece of machinery and we're going to sell it to a customer.'

Now the French were much clearer in their minds that they had to carry the French people with them. The first roll-out at Toulouse was a huge international event. The roll-out of 002 at Bristol took place on a day when I was in the City and I discovered when I got back to my office that they had not even bothered to tell me. I had a tremendous job to get BAC to send 002 to the Paris Air Show. They said: 'Oh, it'll interfere with our test programme.'

When I went to China, which I did in 1971 as an Opposition Member, I raised the matter of Concorde with one of the Chinese ministers. 'Oh yes,' he said, 'the French plane. We are thinking of buying some of them.' So I told him that the British built it too. 'Oh, do you?' he said. 'I didn't know that.'

I think there has been failure in Britain to see the problem in a broad way. I was never quite sure in my discussions with BAC and Rolls-Royce whether they were aware that they were producing something that had to go the environmental lobbies and be accepted by Congress and the Port of New York. It is no good sitting in an office and then complaining that there are political pressures against you. There are always political issues where you have a project of this magnitude.

There is, of course, a certain amount of truth in what Tony Benn has to say about the lack of realization of political forces at work on the part of BAC. Perhaps this is because in developing commercial airplanes we do not concentrate our activity on the Press and the general public. We are not selling articles that retail in the shops, we are selling costly capital goods to airlines. We therefore concentrate our arguments, attention and specifications on our prospective purchasers rather than on the public and politicians. It was not until 1964, with the Brown Paper, that BAC paid any real attention to the political aspects of what they were doing. We preferred to get on with the business of making an airplane and then concentrate on the tough job of selling it to our end customers, the airlines, who had to operate it.

13 Taming the Winds

Concorde is a completely new phenomenon. Too many people regard it from an outmoded, old-fashioned stance. Problems of cost, depreciation, load, purchase price, tax, duty, noise, cost of maintenance and operation must all be looked at in a new light and must be assessed anew. It is no good applying old yardsticks to such a novel concept.

Of course it will have teething troubles, and already in this book I have pointed out some of the difficulties it has had to overcome, but let us not be filled with doom, gloom and alarm like our critics. Instead let us be clear-sighted and look objectively for a moment at what Concorde has to offer its passengers, for it has undeniable assets: shorter time spent in flight means less discomfort and boredom; by going there and back across the Atlantic in a day the biological timeclock is less disturbed and so businessmen will not be so adversely physically affected by time changes and will not arrive for tough board meetings in what is to them the middle of the night; Concorde can carry a great many passengers in first-class style; it makes better use of airport facilities at off-peak hours; it is so powerful that it is not affected by bad weather, so its flight timing is accurate and its arrival time punctual. I shall now proceed to spell out some of these important points in more detail, in order to give some idea of the positive aspects of Concorde in service.

It has always been BAC's contention, now adjusted to the effects of inflation, that, with a first-class fare on Concorde plus a surcharge of between ten and fifteen per cent, the airplane will break even on the North Atlantic with a load factor of something just over sixty per cent, which for a 104/106-seat airplane is a very reasonable figure. The problem is somewhat more complicated than that. If an airline has a fleet of airplanes already carrying first-class, economy and charter passengers, then the people who are making the biggest single contribution to the revenue are the first-class passengers. It does not cost any more to carry them. They occupy a little more space and have rather more expensive food and drink, but not inordinately so, and they pay immeasurably more than block-booked charter passengers.

I believe that our differences of opinion with British Airways in 1974 were a result of them saying not so much: 'Concorde is an unprofitable vehicle' but rather: 'Given the fleet of airplanes that we have already, if you put Concorde on top of it, the net effect across the board may be that although we scoop the pool of the first-class traffic, we may be slightly less profitable than we were before we started.' The other point of disagreement between us concerned the load factor. British Airways' load forecasts have always been a good deal lower than BAC's. That is not a matter of fact. It can only be determined once the aircraft gets into service. BAC thinks its estimates are correct, and time will show.

Henri Ziegler, always the enthusiast on the French side, has never been in doubt that the load factor will be high. He makes the valid point that it is still going to take two or three years to introduce the first sixteen Concordes into service and that their novelty value alone will attract passengers without the dire results predicted by the airlines for their subsonic jets, which will still continue to provide the mass transportation facilities for some years to come. The fare surcharge is also a variable factor and could well be different on some routes from others, depending on their frequency and on the density of occupancy.

The next bone of contention is the purchase price and the cost of maintenance and operation. Concorde now costs about five times as much to buy as it did when the first estimate was made back in 1962. The current figure is of the order of $65m. But then the cost of everything else has gone up in the same proportion, and it is not Concorde alone which has suffered from the effects of inflation. Even so, the British Airways' estimate of operational costs is substantially higher than that consistently put forward by BAC. The difference is partly due to quite a wide disparity in the calculation of the amount of flying time each aircraft can clock up in the year. British Airways have been allowing for Concorde to use something in the order of 2800 hours a year. It is generally conceded in the airline business that a long-haul airplane, which Concorde is, as opposed to a short-haul airplane like the BAC One-Eleven or McDonnell Douglas DC9, ought to be operated for a minimum of 3000 hours a year and really up to something more like 4000 hours. In calculating the costs of operation, it makes a very substantial difference to talk in terms of 2800 hours a year rather than 3500–4000 hours a year.

There is then a difference of opinion about the period of depreciation to be applied to Concorde. Most airlines in the world today depreciate their Boeing 747s over sixteen years. Certainly none depreciate them over less than twelve years. In the case of Concorde, British Airways nominated eight and a half years as the depreciation period for the aircraft, which, combined with the 2800 hours of flying time, increased the cost per hour or per mile or per seat by a very substantial margin.

This second point has been a very curious bone of contention – and British Airways has since been persuaded to adopt a ten-year depreciation period. BAC gives Concorde a guarantee of structural integrity equivalent to ten years at 3000 hours a year. That does not mean you throw the plane away at the end of ten years, only that you might then need a structural programme to prolong its life. I would even accept ten years as a basis for calculation. British Airways thought first in terms of eight and a half, as I understand it, because they had originally assumed that there would be a larger, more advanced and more economic American supersonic transport coming along which would shorten the life of Concorde. That is no longer the case.

BAC had very much the same sort of discussion with BOAC when the airline was preparing its cost estimates before the introduction of the VC10. BOAC first placed a lower depreciation period on the VC10 than we felt was warranted, and in practice, after the airplane went into service, the depreciation period was increased.

As far as resuscitating United States sales is concerned, we are fast approaching the time when the performance, economics and passenger appeal of the aircraft will be demonstrated in passenger service.

On the other hand, Braniff, which operates an extensive Latin American network out of the United States, are convinced that there is a market for the plane, that the economics are right and that the performance is right. Braniff is a very efficiently run, very profitable airline at a time when many of the other carriers are in trouble. They have a very good fare structure in Latin America. They believe Concorde would do a lot for them, so their tendency, as good businessmen, is to say: 'Okay, offer us a deal that we can't refuse.' They do not want to take the risk of damaging the profitable position they hold, but they would lease Concorde on good terms on condition that they could hand it back if the operation did not prove profitable. That is an entirely practical approach. Perhaps neither the British nor the French have worked hard enough publicly to promote the airplane as a profitable vehicle.

Concorde halves the time of air travel. Our critics are at pains to point out that flight time is only a portion of journey time and that delays in ground travel at the cities at either end diminish the advantage. It is not a very persuasive argument. We might just as well have stuck with Douglas DC3s and Lockheed Electras instead of using jet airplanes to fly to Paris or Rome. In practice, all the airlines found that as soon as they put jets on to those routes people flocked to fly in them. The operators found that they not only increased their share of the traffic when they were first with jets, but they increased their total traffic on those routes. Obviously air travel would become even more popular if you could get to the airport more quickly and if the time to check in could be reduced, but it does not affect the overall argument.

To fly to Teheran or the American eastern seaboard in half the time it takes at present makes the most enormous difference to the way you feel when you get to the other end. To put it quite simply, if you get into a motor car and sit in it for six and a half, or seven hours, you feel worn out at the end of the journey. Going right back to the days of stage coaches, the ideal stage was two or three hours.

To add on the time taken in ground transportation at each end is a piece of sophistry. People are concerned with how long they are going to be in the airplane. If you are travelling to Cairo or Singapore or Johannesburg, you do not think in terms of how long the total journey is going to be from door to door. You or your secretary look up the ABC or speak to a travel agent and, if there is one flight that is direct and nonstop and does it in x hours, and there is another one that is either not direct or not nonstop and does it in x plus four hours, there is no doubt which flight you will take. This is the appeal of Concorde.

Another adverse comment that is made is that the internal dimensions of the aircraft are not greatly superior to that of the original Douglas DC3. In fact, the diameter of the cabin is almost precisely that of the Convair 440 which was the next generation in line from the DC3 and the fuselage is of course more than three times as long. The very early mock-ups did tend to have standard hat-racks and looked perhaps a little cramped compared with the jumbo jets. However, all credit to British Airways and Air France who have, with Aérospatiale and BAC, spent a good deal of time and money in devising a new interior which takes full advantage of all the modern techniques and gives a 'wide body' image, moulding the closed-in hat-racks to the shape of the fuselage and giving a good deal more headroom.

There is now plenty of headroom and leg space, although

the seats are a little narrower than those in the Boeing 747. You are only sitting in your seat for an average of three and a half hours, and on most sectors for two and a half hours, whereas the passenger in a long-haul subsonic airplane is often sitting for a minimum of six or seven hours and sometimes as much as ten. Nor must it be forgotten that a number of business people who travel these days, particularly in America, fly in executive jets where the headroom is lower than that in normal commercial airlines.

Our critics, who are prepared to pick up almost any stick to beat us with, have even suggested that there is not enough galley space to maintain first-class service. This is absolute nonsense. Both Air France and British Airways have now shown that they will be able to serve full-scale first-class meals. If there is one point of substance, it is probably that the toilets are slightly more restricted than in a wide-bodied subsonic jet, particularly if you are a male and standing up. I am well over six foot myself and cannot claim to have had any difficulty.

In an actual flight, the first thing you notice is that you get into and out of the aircraft much more quickly than the train-load of passengers carried by a wide-bodied jet. The most extraordinary thing about the flight itself, for what is a completely new experience in air travel, is its normality. You do get the impression of a pretty rapid climb-out, and the angle of the floor in the cabin appears to be steeper than usual. Passage through the sound barrier passes almost unnoticed and the flight is remarkably quiet. The delta wing shields most of the engine noise which, at supersonic speeds, is being left behind anyway. There is a very slight movement, caused by the after-burners coming on, as if you were accelerating in a car. You get a very slight push from the back of your seat. It is barely noticeable and the passengers I have flown with seldom remark on it. The windows are smaller than they are in other jets because they are designed to meet the Federal Aviation Administration's decompression regulations. Apart from that, everything is absolutely normal. You do come over the fence on landing at a slightly higher speed than today's subsonic jets, but this is perfectly acceptable and touchdown is smooth.

It is only in taxiing that passengers might notice a slight difference. It is a sports car ride rather than a Rolls-Royce ride. The airplane responds somewhat harshly to imperfections in the runway. Concorde has a rather long fuselage projecting beyond the nosewheel, which is itself fairly flexible. There is a slight wavy and bumpy motion depending more than anything else on the condition of the runway. In fact Concorde is only going to use the world's major airports, and in these the international regulations covering the condition of the runways are strictly respected. It is not as if it is going to land in every small town in the world.

The design of the main landing gear is a French responsibility, and with its present geometry, it is a little difficult to improve the springing and damping. It can be done, and we and the French have been considering a modification for some time. The work is in train and the modifications will be made to later models.

One enormous advantage Concorde will have is that it will avoid congestion at the popular take-off and landing times which affect subsonic planes flying the Atlantic. The 'stacking' problem is particularly serious in the United States by reason of the curfews imposed in Europe on times of landing at international airports. There was one occasion when I was flying in a jumbo jet when we were the thirty-ninth airplane in the queue waiting to take off. They have to do that because they all take about seven hours to get to London. But in Concorde, which will do the trip in three and a half hours, you avoid the queue. You can take off when the airport is not so busy. You spread the load at the airports and make better use of the ground staff and the airfield facilities. You can leave New York two hours after the last subsonic has gone, and get to the other end an hour before the first one arrives.

Another advantage lies in the cruising altitude. At 35,000 feet, where the subsonic planes fly, the wind speeds are at their most variable and reach their highest velocity. This wind can average on an Atlantic crossing between 60 and 100 mph. If you are flying at 500 mph with a 100 mph wind behind you, and the other way with a 100 mph wind against you, the time of crossing varies by an hour or more. Not only that, if the winds are variable your time of arrival becomes flexible. You simply do not know within fifteen minutes or half an hour what time you are going to get to your destination.

At 55,000–60,000 feet, where a supersonic airplane flies, the winds are only half the velocity, so the effect of a 50 mph wind on an aircraft cruising at 1350 mph is infinitely less. This was illustrated on one of the demonstration return flights between Boston and Paris where, in the one direction, the flight time was 3·08 and the other 3·09 hours. There was literally only one minute difference. This means that you can predict your time of arrival with vastly greater accuracy.

There remains the vexed question at the time of writing of whether we get Concorde into New York at all. Henri Ziegler always took a very pragmatic approach to this. He was against fighting the Americans head-on, and preferred a wait-and-see policy – open a route here, a route there and, as people begin to see the advantage of the service, others will join in. If New York remains obdurate, the services to Rio and the Middle and Far East will prove successful. The American FAA has issued a preliminary Environmental Impact Statement which, at the time of writing, is still being debated. The Authority then has to issue a final statement after it has considered all the objections, and there is a very real prospect that this could be taken to the courts and challenged, a process that might take up to twelve months. That introduces a whole package of massive international complications.

Before British Airways and Air France can extend service with Concorde, a number of measures have to be taken by the British and French governments. On certain routes, supersonic corridors have to be agreed with the territorial powers concerned. Landing rights in places like the United States have to be granted. Over-flying rights in Russia have to be settled and ways have to be found for the Concorde to land for refuelling at a suitably convenient airport in Africa, because it needs to make one stop on the journey from London to Johannesburg.

The French so far have got full clearance to operate Concorde from Paris to Rio de Janeiro via Dakar. At the time of writing, neither Air France nor British Airways have rights into New York nor indeed into the United States. British Airways has inaugurated its Concorde service by flying to Bahrain, after getting certain supersonic corridors in the Middle East to achieve that. These then need to be extended across India if the service is to continue on to Singapore and Australia. A great deal of activity is needed to obtain these rights, to ensure that British Airways and Air France are to have worthwhile routes on which to operate the aircraft. These are problems which I am certain will be sorted out successfully, but I would like to see a greater sense of urgency.

I am often asked what I would recommend as the next stage in the development of Concorde. There are two main lines to pursue. The airplane could be exploited to its maximum supersonic advantage if it had slightly greater range and could therefore avoid the necessity for some of the fuelling stops on its present planned world routes. It is a genuine North Atlantic airplane, but it is not, for instance, like a DC10-30, a genuine London–California airplane, or anything remotely like it. Nor can it undertake the sort of flight of which certain versions of Boeing 747 are capable, such as London–Johannesburg non-stop. Extending the range is therefore the first priority.

Secondly, of course, all of us would like to see more work done to reduce airport noise levels, not only so that the environmentalists have no complaint on that score but because we, in common with most people in industry today, are conscious of our obligations to meet legitimate public demands for an improved quality of life.

I am often asked if we will enlarge Concorde so that it can carry more passengers in the way the subsonic jets have been expanded to jumbo size. My answer to that is no. I do not see it as a necessary or desirable development at this time. Concorde is different from the big subsonic jets. It is catering for one particular section of the market – high-yield, full fare paying, first-class business travel. That is a relatively small part of the market in percentage terms, although it is far and away the most rewarding part, which is one of the reasons why we have always contended that Concorde can be operated profitably. There is no requirement on the basis of present statistics for a stretched or bigger passenger capacity airplane.

At the moment there is no desire to embark on further development of Concorde. In this country, and I think the same applies to France, the attitude is: 'Well, we've spent a thousand million pounds between us on this development, and until we see how the airplane performs in service, whether we sell any more or whether it is a profitable airplane to operate, we are not going to agree another five hundred million pounds on its development.'

It has been an exceptional development project in many ways. We have made the first supersonic transport that the world has yet seen. It works perfectly and meets all the specifications ever written into it. It seems a highly feasible proposition. It is now established in service but unlike any other aircraft I can think of, there does not seem to be anything coming up behind it, largely because of the enormous costs involved in developing airplanes of this type. It is rather like the major space projects. You are limited by the extent to which you can commit national resources to this kind of work. But it is strange that nowhere in the world is there any development follow-up, bearing in mind that the present design was pretty well frozen about ten years ago.

Now, in the mid-1970s, if we had to start again with a clean sheet, we could build a much better, quieter, longer-range airplane. We know how to do it all right. What we do not know is how to get another £1000m, or whatever the figure would be today with inflation, to get such a project off the ground. And at the moment it does not look as if the Americans have changed their minds about the decision not to go ahead with their SST.

Opposition in the United States did not all come from the same source. It coalesced when the environmentalists and those who did not believe that government money should be put into commercial projects found they had common ground. Americans who were in favour of their SST, who included many in successive administrations and most people in the aviation business, never thought for a moment that the anti-SST lobby could possibly win. But they collided with a sudden revulsion against advanced technology because it appeared to lead, or was made to appear to lead, to pollution of the environment.

What I do find extraordinarily difficult to judge is what the true reaction of the famous 'man in the street' is. I get the impression that he is rather pleased that Britain has produced Concorde. Certainly, when you arrive in Teheran, Bahrain or Rio de Janeiro you find the people in these places looking at you with new respect. They knew about American space shots but now, suddenly, they have visible evidence of a standard of advanced technology in Britain and France which they had not expected. Whether that practical benefit to our image has penetrated to the ordinary citizen in Britain, I would not claim to know.

What does seem to me to be a recent development is that on all sorts of subjects and in all sorts of spheres, minorities in Britain have become so vocal that the majority point of view goes by default. These relatively small minorities put themselves forward as spokesmen for the people at large when they are nothing of the kind. Perhaps our public relations on Concorde have been defective. We have rather tended to sit and let the waves of criticism crash over our heads without responding positively. In France, it has to be said, with the closer cooperation and more direct link between Aérospatiale and the government, they took much more positive steps to ensure that majority opinion found proper expression. The turnaround could come. I am reminded of the barrage of opposition that went up from the environmental lobbies about the oil pipeline in Alaska: when the national need for it was established, the agitation was stilled and opposition disappeared almost overnight.

Certainly one supporter in Britain remains sanguine. Julian Amery, whose signature is on the original agreement with the French, has never wavered in his enthusiasm. 'It will probably do more than anything else to save British and European aviation,' he says, 'and it will, for the next decade if not more, be in a class of its own. I think it will pick up an enormous amount of traffic. I thought the Shah of Iran put it rather well when he landed in Australia, having flown in an ordinary subsonic airplane, and when the Australians asked him at the airport: "Why are you buying Concorde?" he answered: "Anybody would who has just made the journey from Iran to Australia subsonic."'

I agree with him that business executives and high officials making these long journeys will be delighted to travel in Concorde even if it involves paying a small surcharge. What is more important in the long run is that we shall have experience of supersonic commercial flying which no other country will be able to match, except possibly the Russians, and they are simply not commercial competitors in the West at present.

We have learned an immense amount about the actual technique and expertise of supersonic civil flight. If we can get the economics right, it should enable us to continue ahead of the Americans when the next generation of even faster aircraft comes along. If we had not built Concorde, I consider that we would have been out of the long-range passenger aircraft business altogether.

14 Bristol at Bay

There is a new dark cloud appearing on the horizon, affecting the whole future not only of Concorde but of the British and indeed European aircraft industry as a whole. Will we be required to manufacture more Concordes over and above the sixteen on the production lines at Bristol and Toulouse, and if not, what are we going to build and how are our people going to be employed? I need to make the point again that building aircraft is a very long-term business. You have to think in terms of ten years ahead. Politics is a short-term business, and this has made for difficult relationships between government and heavy industry, which is what we are.

The outlook for the British airframe industry if we sell no more Concordes is bleak. We are restarting the BAC One-Eleven production line, but the numbers involved are minimal compared with the overall problems of keeping Bristol busy. The Romanians have ordered five One-Elevens, so we have decided to lay down another ten. At least we are making a profit on them. The jigs and tools have long since been amortized, the development costs have been recovered and, from the commercial point of view, the deal is just acceptable. But this, again, mainly benefits the manual workers involved in the manufacture of the aircraft. What we do about our design, development and flight test staffs, who are vital to us if we are going to preserve a capability for the future, I do not at the moment know. You cannot employ them indefinitely with nothing to do, and if you cease to employ them you lose the capability of developing new airplanes, and that is the end of the road as far as a business like ours is concerned.

There is one European civil project going ahead, although it is getting off to a slow start, and that is the A300 European Airbus. BAC is not involved in it. Hawker Siddeley are making the wings as a private venture, but this only represents about fifteen per cent of the cost of the aircraft, and it is hardly a major airframe project for a United Kingdom manufacturer. France has a twenty-five per cent share, the Germans about the same and the Americans are building the engines and supplying some of the equipment.

If the Airbus attracts the attention of the airlines there could be a Rolls-Royce engine in the later versions, but that still lies in the future. It is a combined project under the control of a single management at Toulouse, bringing in people from the other countries involved as part of the team. The sales manager is a Dutchman and Hawker Siddeley have some of their people on the sales side. This is a spin-off from the Concorde experience – where at least we learned how not to organize these things – and provides a useful pointer to the way international cooperation should be organized over the next decade.

The outlook for the engine side of the British aircraft industry is much brighter. In world terms there are only three engine manufacturers: Pratt & Whitney and General Electric in the United States, and Rolls-Royce in Britain. Each of them can be regarded as on equal terms competitively. The Bristol Engine Division of Rolls-Royce has plenty of work in hand: even with the present limitation on Concorde they are in process of making sixty engines for the development programme plus another ninety for the production models and airline spares. Even in its present sophisticated form, the Concorde engine represents the final development of a design that was first run twenty-five years ago, and it has about reached the end of the road. However they have all the knowhow and experience to produce a second-generation model the moment the requirement arises.

Rolls-Royce at Bristol has triumphantly survived the

bankruptcy of the parent company at Derby, which went into liquidation over the disastrous contract they had signed with Lockheed for the RB211 engines which now power the 1011. It goes to show how little the component firms were integrated. At Bristol they are making Pegasus engines for the Harrier, the Viper for numerous smaller jets, and the RB199 engine for the new European Multi-Role Combat Aircraft. This is a major project. The British, German and Italian governments are expected to buy a combined total of 2500 engines, which represents about £750m worth of business. It is a tremendous sum of money. Production will go on for a number of years, and Rolls-Royce could well end up selling £2000m worth of the RB199 engines alone, which will keep them going at Bristol for a long time.

Moreover the Olympus engine provides the take-off point for all manner of future development. Applied to industrial marine work, it will generate forty-five or fifty megawatts of electricity in one unit. It gives around 50,000 horsepower for driving industrial machinery in ships. It is the most powerful engine in the world from that point of view. If the next stage in supersonic flight is an airliner weighing 700,000 or 800,000 lb, it will be necessary to put a fan in front of the current Olympus engine. The Americans, despite the expenditure on their abortive SST, have nothing to compare with it. Rolls-Royce has done an enormous amount of compressor and turbine work, and in those fields their experience is pre-eminent. They also know a great deal about materials, lubricants, and the testing of components for engines. It has put Rolls-Royce at Bristol at the forefront of the world in turbine engine technology.

That, however, does not solve matters on the airframe side. Concorde number sixteen, as things stand, may well be the last in the line. We were completing the first components for it in 1975, and progressively parts of the factory are running out of work. Unless something is done immediately, the only people left with something to do will be a small group of assembly workers putting together the last model, which should fly from Filton in 1978. Of course the jigs and tools can be kept in cold storage. We did this with the BAC One-Eleven when there was a big gap between the last plane made by the main production line and reassembling them to meet repeat orders we received from Romania, but even so it is not practical to keep the jigs in being for more than three or four years.

You would think that there would be the most frantic dialogues going on between the French and the British governments about what they are going to do now. After spending a thousand million pounds, there seems to be no programme or plan of any sort being evolved. I know of no discussions going on at all. There seems to be a Micawberish idea going around that something will turn up, but unless it happens very quickly it is not going to make very much sense. The unions are vocal enough when it comes to wage demands and you would think that they would be banging on the government's door asking them to do something about this situation. Nothing is happening and the situation gets more serious every week.

A break in the production process of Concorde would be additionally damaging because there is always the risk that the environmentalists will take up the cudgels again with a campaign to prevent starting up production lines in 1980. They might well win their battle the second time round, and the tools for building Concorde would be so much scrap.

Sales negotiations are going on the whole time, particularly in the Middle East where all the oil money is, but it is no good pretending that there is a long queue of people fighting to lodge further orders. Foreign airlines are waiting until the aircraft is proved in service and supersonic corridors and landing rights are cleared for the main routes. They will then want to see if Air France and British Airways succeed in operating it profitably. They assume it will remain in production and that they can start picking the airplanes up whenever they feel like it. Right now it is difficult to think of a good carrot to entice people to commit themselves to the plane before it goes into service.

The moment more orders start coming in, we would have a very good case indeed for saying to the government: 'You know, Concorde has started to be the success we always said it would be. We must start building some more.' It really would be frustrating if, when the first sixteen aircraft go into service, they are a wild success – full of passengers, and profitable – but we had no production beyond that to meet demand from airlines. That would be the ultimate piece of nonsense after spending a thousand million pounds. The Americans are not within years of us, so they would not benefit and so the situation might arise for the first time ever where the only place to go and buy new airplanes would be the Soviet Union.

In the first instance, even a major world airline would probably not need to buy more than half a dozen Concordes. The plane is not going to cater for the mass travel of tourists and charters, it is going to attract the top ten to fifteen per cent of high-yield traffic. As things stand, it would probably suit the major world carriers best if Concorde were to go away and never come into service. Most of them are in very poor shape financially. Pan American is involved in disastrous losses and TWA is in serious trouble. The last thing they want is this new animal getting into the circus ring and changing the whole act.

Their dilemma is serious. With the losses they are making, they do not want to pay out $60m-odd a time for this piece of new equipment, but on the other hand their losses may get worse because some of their prime high-yield traffic may be taken away by Air France or British Airways. It could be a very difficult decision for them to take, and most of them are adopting a 'wait and see' attitude. If the business they lose is less than they feared, they will be able to settle down again to operating their 747s, which all the airlines have, and they will all be in the same boat together. So they are sticking their heads in the sand and hoping that the storm will blow over.

One of the reasons why Concorde is having such an uphill fight is the way the world's aircraft industry has developed since the war. Europe represents about twenty-five per cent of the market in the Western world, although it is going up.

The rest of the world, outside America, probably absorbs about the same amount, but fifty per cent of all the civil airplanes in existence are operated by United States carriers. The most important statistic of all is that they have 93·5 per cent of the production. There is nothing wrong with European technical standards and capability. We are certainly the equal of the Americans. Where they do have the edge is in the sheer volume of their production and market, and the force of their sales drive.

The two main world suppliers of airplanes are McDonnell Douglas and Boeing. Each of them sells every year massively more than the whole of the rest of the manufacturers around the world put together. The Americans have, since the war, been putting on to the world market consistently good, rugged, serviceable and economic airplanes, although there have been some exceptions. They have always maintained a continuous line of activity, with models succeeding one another over the years. You cannot succeed in this business in fits and starts – make an airplane, finish it, have nothing else for the moment, pack up, go home and come back again in five years time. I was talking recently to the President of an American company who told me that part of Lockheed's problem today has been the difficulty of putting together a team for the Lockheed 1011 after the long period from the end of the Electra when Lockheed were producing no commercial airplanes at all.

The Americans make a great virtue out of their claim to develop their planes on a commercial basis. This is only partly true, and one factor always needs to be borne in mind when comparing the Concorde experience. American manufacturers have received enormous help in the way of government finance for their military and defence projects. The General Electric engine that powers the DC10, for example, owes a great deal to half a billion dollars' worth of military development of the engine that initially went into the military C5A. The Boeing 707 owed much to the military tanker which preceded it. Until the Lockheed 1011 got into serious trouble, the Americans had not been prepared to finance directly the development of civil airplanes, and the opposition that mounted to the expenditure of government money on developing their SST was one of the principal causes of its cancellation.

It is extremely tough to sell into the United States. Apart from anything else, there has been and still is an import duty imposed on European airplanes. Strangely enough, although in theory there is an import duty on American planes coming into Britain, it is very seldom paid because there is a provision whereby the Department of Trade can allow a waiver of duty. In America there is no such waiver. The duty always has to be paid. Currently it is running at five per cent, but a lot of our BAC One-Elevens were sold when it was running at twelve and a half per cent, and when they took the Viscounts it was between fifteen and seventeen per cent. So that duty has to be added to the cost to them of Concorde.

Around five thousand people are still employed on Concorde at the Filton works of BAC, and the work is running down. It is difficult to see how you can put work into the plant quickly enough to preserve the jobs of the people there. We are going to find it progressively more difficult to provide work for anything like the number of people at present employed at the Filton works unless, as a matter of urgency, more work is injected now, not just minor penny packets but something of consequence.

In the rest of BAC, there are several thousands of people, mainly at Weybridge and Hurn, also employed on Concorde work, and this represents a pressing problem. The effect on the rest of the industry would be more marginal because they are mostly equipment suppliers and subcontractors, whose workload varies in percentage according to the Concorde content, but they have other outlets and might be less affected. Even at Toulouse, where they have the A300, the effect would be pretty devastating. There is no major civil aircraft project coming along and the situation on the military side is fairly static.

I do not even see how we can introduce the sort of solution that tides over the motor car industry when it gets into difficulties, that of putting our people on short time. An aircraft is a different product to an automobile and cannot be turned out in a single day. I have read that what sometimes happens in the motor industry is that the people are put on a three- or four-day week, with a little overtime, and then laid off for the other two days, which entitles them to go along and collect two days' social security benefits, and they end up probably not much worse off than they would be if they were working a full week. But this would not be a practical solution to the Concorde problem.

If sixteen production aircraft end up forming the total, people will be entitled to ask what on earth the French and British have spent a thousand million pounds for. The expensive jigs and tools would only have scrap metal value, and if we do no further work on the development of supersonic transports it will end up as an interesting but extremely expensive experiment.

The effect on Bristol itself will be devastating. We have always had very good labour relations at Filton, apart from a few minor stoppages, and we have always been able to maintain something of a family atmosphere in the works and a high degree of regional loyalty. Concorde, with its advanced technological demands, has been a source of immense pride to our people. Even after the merger there has been a very close feeling of unity among the people in the Filton works at all levels, combined with pride in the project on which they are working. If no more Concordes are to be built, all this would go. The dispersal of the real and unusual talent available there would be a disaster.

Most of the fitters and assembly workers, with their acquired skills, would probably find useful work elsewhere. The people who would be irreplaceable if they were dispersed would be the hard core of specialist people who work in areas such as design, development, sales and so on. There are not all that many of them, probably not more than a few score all told.

But if they go, a number of other experts in aircraft furnishings, the test flying of development airplanes, and the hundred and one other different specialities that go into a new model, will not have anything to do either. If the British aircraft industry cannot develop another commercial airplane project very soon, we may not find ourselves able to do it at all. We would have the most enormous difficulty in recruiting and training people all over again because we would have lost our basic capability.

The ideal solution to make use of all this knowhow would be to go on building better forms of Concorde. We know how to get significant improvements on the noise problem. They are fairly costly and we would have to see a bigger market for the plane than we have before anyone would be prepared to put up the money. We know how to improve range and reduce noise at the same time with one set of modifications, and we know how to reduce noise without range with another set. An engineer like Mick Wilde will say that this is what the country should do if it can afford to do anything. He would like to see the experience fed into supersonic flight rather than to try to find minor applications of the expertise in such things as business executive jets. He is one of our optimists, I am pleased to say, and he is absolutely convinced that Concorde will sell.

It is perhaps pertinent to look for a moment at what the possible civil aircraft alternatives might be in terms of future development if Concorde continues at its present standstill. If I were asked to put my money on the next runner in this very expensive horse race, I would say that the airplane to go for would be a two- or three-engined, 180-seater, with minimal fuel consumption, quiet noise levels, probably with a wide cabin and with optimum economics at between 800 and 2000 miles range. That way you can meet the short-haul requirement but also do the longer-haul inclusive tour journeys to places like Las Palmas. It is an airplane for which there is a market if the price is right.

The aircraft I have in mind would be subsonic, flying at something like Mach 0·8. Boeing were on the point of producing an airplane to meet this requirement with a further stretched version of their 727 to be called the 727-300, with new Pratt & Whitney engines. United Air Lines were going to put in an order for it, but with the general downturn in business they preferred to postpone their decision and the market is still open. McDonnell Douglas are also known to be studying this requirement very closely at the present time with a twin-engined version of their DC10, smaller in size and with a new wing more suitable to a smaller twin airplane.

The new airplane for this market would require a supercritical wing which would have the effect of improving both its cruising and airfield performance. Great care would need to be taken to ensure maximum fuel economy in view of the continuing enormous increases in the cost of oil, and, of course, airfield noise would have to be reduced to a minimum.

Helicopters are not part of the next stage in major commercial airline travel. There is no way you can operate a helicopter carrying commercial passengers and make money at today's fare levels. They are expensive animals to operate. You can make money with them on contract jobs, like servicing oil rigs, where it definitely pays to be able to lift goods and people into areas not reached easily by other means of transport. You pay a high price in terms of weight and sophisticated equipment to get the vertical take-off and landing capability.

Another possibility, which derives from one of the early alternative designs for Concorde, is the flying wing. This is best applied to cargo planes. There is a strong case for producing a rugged, slightly agricultural piece of equipment, low on fuel consumption and not very fast, that will carry large amounts of cargo from one place to another. Nearly all the cargo planes that are operating around the world are simply versions of passenger airplanes. No company has yet managed to make a business out of building an airplane specifically for freight. I am not sure why this is and I would think it merits study.

The essential difference is that a freightplane, or should I say a bulk-carrying freightplane, does not require pressurization. If you can do without that, it is an enormous saving. It does have to fly at lower altitudes, but as there will be no passengers on board they would not be inconvenienced by the weather conditions, although the flight crew might. Perishable freight tends to be light and has to get to its destination quickly, and therefore that would still need to be carried around in the present converted passenger planes or in the pressurized cargo holds of airliners. In terms of aerodynamics, the flying wing could have much of the Concorde experience applied to it, but its engine would of course be subsonic, and Olympus and its derivatives would not apply.

In this superficial summary of the prospects of the British aircraft industry, there are one or two more points to be clarified. I have described the effect that the non-continuance of Concorde would have on the work force at Bristol. In overall terms, related to our manufacturing and development activity, this would not affect the survival nor dramatically affect the profitability of BAC as a company. We are broadly enough based to survive. The aircraft industry runs in cycles. You go through periods when you have a substantial number of civil aircraft in production, providing a useful chunk of the profits at a time when the military contracts are in the stage of development and therefore not providing much return in terms of work and profits. You then get that position reversed, with either military aircraft or guided weapons providing the profits and your civil aircraft programme running down.

As things stand, BAC is in the second position. Our guided weapons and military aircraft divisions will be providing the greater bulk of our profits for the next few years, mostly from the export business, and our civil aircraft programmes will be on the decline. On the military side we have an order book of some £800m–£900m, with export orders totalling £600m of that. Although the end of the Concorde programme would cause us grave industrial problems at Filton, this would make very little difference to the financial picture of BAC. That is a separate problem from the future of the British aircraft enterprise as a whole. There are effectively only two manufacturers of any consequence left, Hawker Siddeley Aviation and ourselves. Neither has a new commercial aircraft programme coming up, and even if we dreamed one up tomorrow it would probably be about eighteen months before we were able to launch it into development and another four years or so before we had production airplanes available for delivery. And that, in terms of long-range prospects, is the most worrying aspect of all.

While Tony Benn was at the Department of Industry, he took the initiative in bringing together Sir George Edwards and the management of BAC with the shop stewards not only from Filton but from other BAC sites, to discuss how the present hiatus situation might be resolved. These talks produced no solution, except that it is difficult not to see some link between them and the current proposal of the Labour government to nationalize the aircraft industry some time during 1976.

Nationalization is the next situation that has to be confronted and, so far, there has been virtually no discussion about it with the industry. The main reason advanced for nationalization is a somewhat odd one. It is not that we are lame ducks or that there is no way of finding the capital needed to carry on our business except if we are owned by government. The reason advanced is that we spend a lot of public money – which so far as our defence work and programmes like Concorde are concerned is true – and that there is not a sufficiently high degree of public accountability.

The government's Consultative Document on Public Ownership of the Aircraft Industry that was issued in the spring of 1975 by the Department of Industry only contained about half a dozen pages. The main issue was dismissed in a paragraph. It said in effect that the only way of achieving this greater public accountability was by public ownership of the companies concerned, without examining any alternative. It is not suggested, for instance, to take Concorde as an example, that BAC's accounting and cost control procedures have been more lax because it has been public rather than private shareholders' money that has been spent. Precisely because it was public money we were spending, we have been meticulous about the way it was used, and our profit-margin has been nominal. It might be possible to tighten up the degree of accuracy of forward disclosure of how we are going to spend government funds, but even there the system works remarkably well. As far as the people on the shop floor or in middle management are concerned, I doubt very much if they would notice the change in ownership, providing the enterprise was kept intact. But that is not in itself a reason for nationalization.

BAC is in fact an extremely efficiently run company. Each year we submit to our major shareholders, GEC and Vickers, a budget for the following year. We submit cost returns to them monthly in a particular form devised by GEC to answer all the questions that need to be asked. Providing that we are doing all we said we would do, or better, we are very much left alone to get on and run the business. Whether that would be the same or better under public ownership, is a very questionable point. The record of nationalized industries does not give cause for optimism.

There ought to have been a flaming row about all this when the Bill came up for debate, particularly as the terms of compensation the government has proposed are really a piece of daylight robbery. Shareholders are to be paid on the basis of stock market values during a period prior to February 1974, the argument being that this would be a fair valuation prior to the Labour government returning to power with its declared policy. But stock market values were at a low level during that particular period. The most indefensible element in the calculation is that Hawker Siddeley Aviation and ourselves are not publicly quoted companies. Hawker Siddeley Aviation is a wholly owned subsidiary of the Hawker Siddeley Group, representing about forty per cent, and BAC is wholly and co-equally owned by GEC and Vickers Ltd. The government proposes to pay compensation on the basis of what the quotation would have been if the shares had been quoted during the six months prior to February, so there is no basis for calculation at all, particularly as there is no longer an aircraft sector of the stock market.

Coupled with all this has come a certain amount of agitation from Common Market headquarters in Brussels for the rationalization of all European aircraft production. This may look neat and tidy to the bureaucrats in the EEC, but it would in fact be even more counterproductive than the nationalization proposals in Britain. It would probably kill off the European aircraft industry in a very short period of time. Once there ceases to be any competition, we would be quite incapable of facing up to the Americans and they would simply take over the world market entirely. There are several levels of activity in Europe, from Concorde through the Airbus to executive jets and small planes, and to have these all controlled by one centralized and unprofessional administration would be to defeat the object of diversity.

There would need to be at least two major conglomerates of one sort or another in the commercial airplane manufacturing business, otherwise there could be no design competition, no competitive bids, and the well-known financial evils of monopoly would take over. France has already pared down her sources of aircraft manufacture to an irreducible minimum and so have we in Britain. Both countries have now accepted the necessity for international cooperation in any major future projects, but once the industry becomes bureaucratized it is the end of all enterprise.

Concorde, by trial and error, has provided us with the experience necessary to organize the next international project efficiently. That is a spin-off. It stands out in its own right as the most technologically accomplished product in the history of aviation. Air travel provides the one remaining development area in the field of transport. It is not like trains or ships or cars, where you come up against the physical limitations of the environment. We can travel higher and we can travel faster where the air is thinner and the resistance lower. In spite of the intensive studies that the Americans made for their own supersonic transport, their plans have been put in cold storage. Britain and France are still many years ahead of the field. Neither Boeing nor any other US company could, if they started on an SST again now, have one in service in much under ten years.

The success of Concorde in airline operation remains to be demonstrated, but I have no doubt that our predictions will prove correct, once British Airways have published their first annual accounts after a year of operation with Concorde. The astonishing public reaction to the offer of a free journey on one of our testing flights proves far more than the complaints of the environmentalists. There were something like half a million applications for thirty-five places. People will regard their first flight in Concorde as a worthwhile experience.

The first two airlines concerned have to cope with an adjustment to their schedules and the pattern of their service. There will be battles to fight about the fare structure. Some of them would probably prefer the quiet life involved in flying their 747s and Lockheed Tristars and so on, and sharing the profits on the routes on which they operate in pool with other carriers. But there is a good deal of quiet enthusiasm for Concorde. It is a challenge to those who want to see air travel enter a new generation and want to see it work.

The introduction of a new era in air travel always arouses opposition. When the first subsonic jets started operating, they immediately made obsolete all the propeller-driven airliners which were still flying. There were the same complaints as we have with Concorde, that the jets would prove uneconomic and difficult to operate. Almost overnight they changed the face of air travel, and the immense expansion in the number of passengers and consequent reduction in the price of air tickets was the direct result of the revolution which had been fiercely resisted.

One simile that has been used about Concorde is that it is rather like the Hope diamond – a flawless jewel that has brought bad luck to everyone connected with it. This I must reject. It has acquired a life of its own. It has introduced an entirely new element into the air transport industry. It has also affected the international relationship between governments and aircraft companies in the construction of a civil airplane.

Of course, not all was for the good, and the sheer duplication of resources and effort involved over the years of Concorde's construction was wasteful, but this will not be repeated. The way in which the project was administered entailed a wholly unnecessary degree of bureaucratic interference in what should have been technical and commercial considerations. That lesson has been learned, and succeeding projects will be organized much more rationally.

One element that I do regret is that it has absorbed so much of our time and thinking and effort for so long. There are several thousand people down at Filton who have worked on no other aircraft for more than ten years. This has denied them the breadth of experience they enjoyed in earlier days. Concorde has been all-absorbing and a hard taskmaster. Over the years it has dominated a large part of the lives of many of the people associated with it, and to a considerable extent mine as well.

Technically, Concorde's is a triumphant story. Apart from the American space programme, I can think of no other aircraft project involving high technology that has come through with such success. We were probing the frontiers of knowledge all the time, and advanced the state of our art at every stage. Concorde is at one and the same time a monument to British and French engineering genius and a portent of the future. I hope everyone enjoys flying in it as much as I do.

Index